SPIRITUAL INTELLIGENCE

*Significance, Applications, Measurement, and
Development Techniques*

Prof. (Dr.) Jai Paul Dudeja

Made with ♥ on the Notion Press Platform
www.notionpress.com

CONTENTS

SECTION-I

Spiritual Intelligence: Introduction and Overview

SECTION-II
Skills, Aspects, and Components of Spiritual Intelligence

SECTION-III

Measurement Techniques of Spiritual Intelligence

SECTION-IV

Application and Benefits of Spiritual Intelligence

SECTION-V

Development Techniques for Spiritual Intelligence

PREFACE

Dear Readers,

I am extremely happy to see this book titled, **"Spiritual, Intelligence: Significance, Applications, Measurement, and Development Techniques",** in your hands. It is my firm belief that you have chosen to read this book with a specific aim in mind, and I assure you that you will not be disappointed with it.

There are three types of intelligences: cognitive, emotional, and spiritual, with the associated quotients: Intelligence Quotient (IQ), Emotional Quotient (EQ), and Spiritual Quotient (SQ), respectively. Whereas, the IQ measures the intellectual capacity of a person, EQ measures the person's self-awareness, feeling of empathy to be sensitive to situational circumstances; SQ refers to the persons' ability to build their capacity for the meaning of life, the vision and value they hold, and strive for the self-actualization by connecting with inter self and the higher self. It has interface with spirituality to help oneself create a balance amongst the various facets of one's life. Spiritual Quotient (SQ) is the sum total of IQ and EQ. IQ is the functionality of left brain; EQ is the functionality of right brain; and SQ is the functionality of whole brain. SQ has been understood to be the most fundamental of the three 'Q's.

This book explains the skills, aspects, components of Spiritual Intelligence and how to apply, measure and develop one's SQ.

This book consists of the following **Five Sections, distributed in 20 Chapters:**

SECTION-I: Spiritual Intelligence: Introduction and Overview; (Chapter 1).

SECTION-II: Skills, Aspects, and Components of Spiritual Intelligence; (Chapters 2-15).

SECTION-III: Measurement Techniques of Spiritual Intelligence; (Chapters 16-18).

SECTION-IV: Application and Benefits of Spiritual Intelligence; (Chapter 19); and

SECTION-V: Development Techniques for Spiritual Intelligence (Chapter 20).

Following is the list of **20 chapters**:

Chapter 1:   Introduction and Overview of Spiritual Intelligence

Chapter 2:   Self-Awareness

Chapter 3:   Compassion

Chapter 4:   Ego Death and Ego Dissolution

Chapter 5:   Spiritual Maturity and Stability

Chapter 6:   Spiritual Acceptance and Surrender

Chapter 7:   Forgiveness

Chapter 8:   Spiritual Wisdom

Chapter 9:   Gratitude

The author sincerely believes that a book of this nature will be appreciated by all the readers across the globe who wish to understand the significance of Spiritual Intelligence and apply it in their own life.

The author would open heartedly love to receive any encouraging/ critical comments as well as feedback from the dear readers at his Email ID: drjpdudeja@gmail.com

Sincerely

Prof. (Dr.) Jai Paul Dudeja

2024

ACKNOWLEDGEMENTS

The seeds of my interest in the quest for spiritual wisdom were lovingly sown by my revered parents: **Late (Dr.) Shanti Sawrup Dudeja and (Late) Mrs. Jai Devi Dudeja**. I am sure that they are watching me every moment from wherever they are in the other world and continuously sending their blessings to me.

I have greatly benefitted in going through the books, research papers and articles, referred to in the 'Bibliography' in this Book. I gratefully acknowledge these authors for enhancing my understanding on the subject matter of this book.

My greatest admiration is reserved for **Mrs. Rita Dudeja**, my wife and my best friend. She is a continual source of inspiration for me and is a co-traveller on this path of trust and truth.

2024   **Prof. (Dr.) Jai Paul Dudeja**

PROFILE OF THE AUTHOR

Born in June 1948, Prof. (Dr.) Jai Paul Dudeja holds a brilliant academic record. He did his Master's degree from Birla Institute of Technology and Science (BITS), Pilani (India), and Ph.D. degree from the Indian Institute of Technology (IIT), Delhi. He has been in regular employment as a Scientist, Professor, Dean, Director, Principal and a Senior Administrator in various educational institutions, universities, laboratories, public and private organizations. He superannuated as a Senior Scientist and Additional Director in May 2008 from the Defence Research and Development Organisation (DRDO), Government of India. After DRDO, he served for over 11 years, till his last posting as a Director at Amity University Gurgaon, from where he retired in Nov 2019.

Till date, Dr. Dudeja has published/presented about 90 research papers in various national and international journals/conferences. Out of these, over 20 research papers are on spirituality and consciousness etc. Besides this, he has authored the following **twenty-four books** on spirituality and consciousness:

1. Gayatri Mantra: A GPS to Enlightenment,

2. Maha Mrityunjaya Mantra: An Invincible Armour for Conquering Death

3. Ajapa-Japa Sohum-Humsa Mantra: An Eternal Mantra for Inner Consciousness

23. Near-Death Experience: Scientific Interpretation (Inspired by True Story of a Friend)

24. Universe before the Big Bang: A Deeper Insight into Cosmology

Dr. Dudeja has delivered many invited lectures in international conferences on Spirituality and Consciousness in India and abroad.

He has been initiated to more than half-a-dozen meditation techniques.

Dr. Dudeja has been recognized as the "World's Who's Who in Science & Engineering".

SECTION-I:

Spiritual Intelligence: Introduction and Overview

CHAPTER 1

INTRODUCTION AND OVERVIEW OF SPIRITUAL INTELLGENCE

1.1 What is Spiritual Intelligence?

1.1.1 Intelligence

The beginning of intelligence theory goes back to Plato and Socrates who reasoned that intelligence would always organize things in the best possible way. Darwin and Galton added that human intelligence is evolutionary and contributed to the degree of success people have in life. Different psychologists have formulated various definitions of intelligence. Today, the nature of human intelligence is considered one of the most controversial and highly debatable areas of psychological theory and research.

A summary of the definitions reflect that human intelligence is an evolutionary and developmental capacity that is qualitative and of multiple kinds used for adaption to the environment through assimilation and accommodation. It gives humans the analytical, creative, and practical abilities to live successfully by solving problems, creating products, and delivering outcomes within a specific culture.

1.1.2 Spiritual Intelligence

Spiritual intelligence is the ability to behave with compassion and wisdom while maintaining inner and outer peace regardless of the circumstances. Spiritual Intelligence (SI) or Spiritual Quotient (SQ) is a term to indicate the spiritual parallels with IQ (Intelligence Quotient) and EQ (Emotional Quotient). Spiritual intelligence is consisted of two words: spiritual and intelligence. The word spiritual, derived from Latin word *spiritus*, means "that gives life or vitality to a system". Spiritual intelligence is the highest form of intelligence. It enables us to understand more about the creator within. It moves us from being uneducated to be aware, from being one-dimensional to multi-dimensional, from being disempowered to empowered within.

1.1.3 Difference between Spirituality and Spiritual Intelligence

The definition of spirituality as stated by Oxford Dictionary is "relating to or affecting the human spirit or soul as opposed to material or physical things." Spirituality refers to a broader concept that involves seeking meaning, purpose, and connection to something greater than oneself, such as a higher power, nature, or the universe. It is often associated with religious beliefs and practices, but can also be expressed through other forms of personal or cultural beliefs and values. The basic motivation for the action of the spiritual sphere is the search for the meaning of life.

Spirituality includes (a) focus on ultimate meaning, (b) awareness and development of multiple levels of consciousness, (c) experience of the preciousness and sacredness of life, and (d) transcendence of self into a connected whole"

"Just as candle cannot burn without fire, men cannot live without a spiritual life."

– Buddha

Spiritual intelligence, on the other hand, is a specific aspect of intelligence that involves the ability to access, apply, and integrate spiritual and/or religious values and beliefs into daily life. It involves skills such as self-awareness, empathy, compassion, and moral reasoning, as well as the ability to use spiritual values and beliefs to navigate complex ethical and moral issues.

Spiritual intelligence has been characterized as a distinct set of mental capacities contributing to refined self-awareness, enhanced personal meaning, heightened spiritual stats of consciousness, and virtuous behavior.

Spiritual intelligence is defined as a set of mental capacities, which contribute to the awareness, integration and adaptive application of the nonmaterial and transcendent aspects of one's existence.

In other words, spirituality refers to one's beliefs and values, while spiritual intelligence refers to one's ability to apply those beliefs and values in a practical and meaningful way.

1.2   What is the Difference between IQ (Intelligence Quotient), EQ (Emotional Quotient) and SQ (Spiritual Quotient)?

For the longest time, an individual's **intelligence quotient (IQ)** was thought to be one of the key indicators of leadership potential. This notion assumed that the smartest person would be the best equipped to come up with the most effective cognitive ability and leadership strategy. In the early 20th century, various psychologists had contributed towards devising tests to measure an individual's intelligence, his reasoning and logical powers. It demonstrates the rational intelligence, ability to manage facts and information, and. use logic and analysis to take decisions. IQ is tested using the Stanford-Binet Intelligence Scales. It refers to our rational, logical, rule-bound problem-solving intelligence. It is supposed to be what makes us bright or dim. It is also a style of thinking. All of us use some IQ, or we wouldn't be functional.

EQ refers to our **emotional quotient**. In the mid-1980s, in his book "Emotional Intelligence: Why It Can Matter More Than IQ", Daniel Goleman articulated the kind of intelligence that our heart, or emotions, have. EQ is manifested in trust, empathy, emotional self-awareness and self-control, and the ability to respond appropriately to the emotions of others. The core concept of the study was developed on an individual's feelings, his/her own feelings for others and aptitude to manage the same.

Emotional intelligence is the ability to monitor one's own and others' feelings and emotions, to discriminate among them and to use this information to guide one's own thinking and actions.

Whereas, the IQ measures the intellectual capacity of a person, EQ measures the person's self-awareness, feeling of empathy and their capability to be sensitive to others' feelings, the **SQ (Spiritual Quotient)** refers to a persons' ability to build their capacity for the meaning of life, the vision and value they hold, and understand the elements that make them dream and strive for self-actualization. It has interface with spirituality to help oneself create a balance amongst the various facets of their life.

Spiritual Quotient (SQ) is the sum total of IQ and EQ.

$$SQ = IQ + EQ.$$

IQ is the functionality of left brain; EQ is the functionality of right brain; and SQ is the functionality of whole brain. Without EQ, people cannot effectively use IQ. EQ is the fundamental need for effective use of IQ. SQ includes both intelligences: emotional and intellectual intelligence. SQ is the ability for integrating of inner and outer self of life. Spiritual intelligence is the expression of innate spiritual qualities through our thoughts, actions and attitude. Spiritual intelligence is the ability to access our higher selves and therefore access higher knowledge within ourselves.

However, to tune into this, we need to be able to identify what is our intuition and what is our ego. Through knowing our triggers, releasing our emotional baggage and forming a relationship with and taming our ego, we are able to cut through the noise and tune into this intelligence. Spiritual intelligence can help deepen our sense of awareness about a cognitive growth mindset.

IQ, EQ and SQ correspond to 3 distinct neural arrangements in the brain. In knowing only IQ and EQ, Western psychology effectively places a hole at the centre of the self. Models have had 2 layers: outer, rational, conscious personality, and the inner, unconscious one. Now we have found a third layer, a central core. Conscious personalities can be described with the standard personality profiles.

SQ is not connected to religion. People can be religious and spiritually stunted. More people have religious experiences outside the confines of mainstream religious institutions than within them. SQ is the soul's intelligence, the one with which we make ourselves whole.

There is a difference between being religious and having a high SQ. These should not be confused and used interchangeably. Studies have proved that there is little correlation between religiousness and level of SQ of an individual. Religion is a set of laws and beliefs which govern our society and is external to an individual, whereas SQ is internal intelligence which is governed and developed by oneself to recognize the values which exist within him/ her and discover the new ones.

What differentiates great leaders from merely good ones is the nascent but profound idea of the spiritual quotient, SQ, which essentially reflects the extent of a leader's self-realization. It captures the leader's level of self-awareness and the sense of self-responsibility. What they deeply stand for, their non-negotiable values are, their strength of character and their commitment to a clear personal purpose and its relatability to the mission.

SQ has been understood to be the most fundamental of the three 'Q's. SQ is used by an individual to develop his/her life and understand the meaning it holds for him/her. Comprehension of the factors which make an individual live every-day of his/her life, with meaning and aim for a higher platform of satisfaction and contentment. It calls for the perfect balance to be created.

SQ is the final bastion of leadership growth. When leaders grow their SQ, they operate from a healthy combination of self-confidence and humility, passion and empowerment, and producing results and creating sustainable organizations. While you need raw intelligence and the ability to connect, inspire and influence people, operating from a place of higher self-awareness, conscience and purpose helps you break through to the highest levels of leadership effectiveness.

SQ is the intelligence with which we address and solve problems of meaning and value, the intelligence with which we place our actions and lives in a wider, richer, meaning-giving context, the intelligence with which we can assess that one course of action or one life-path is more meaningful than another. SQ is the necessary foundation for both IQ and EQ; it is our ultimate intelligence. It allows us to be creative, change the rules, alter situations; to dream, aspire, see the uses and limits of both understanding and compassion. EQ allows us to judge a situation and behave appropriately within it; SQ allows us to ask if we want to be in it at all, or would we rather change it, create a new one?

Spiritual intelligence has been defined as the adaptation of spirituality to help one face and solve the every-day problems of life on the personal or professional front. Spiritual intelligence is the expression of innate qualities through your thoughts, actions and attitude.

What makes a great collaboration? One view is that a collaboration is only as great as the individuals who collaborate within it. Another is that a collaboration is also only as great as the vision that drives it. A vision is something you reach for, something you aspire to, something that is the glue of your enterprise, the driving force, the vitality within it. In our world today, the thing we are most lacking is vision.

Part of the reason many of our collaborations lack vision is because they're based on only one kind of capital — material. It's true that any kind of enterprise we want to engage in, requires some kind of financial wealth if it wants to succeed in the short term. But, for a collaboration to sustain itself over the long term, it needs two other forms of wealth: social and spiritual. These three types of capital are connected similarly to a wedding cake. Material capital sits on the top layer, social capital lies in the middle, and spiritual capital rests on the bottom.

Even more fundamental than social capital, spiritual capital reflects what an individual or organization exists for, believes in, aspires to, and takes responsibility for. Based on this definition, it is a new paradigm that requires us to radically change our mindset about the philosophical foundations and practices of business, or any enterprise for that matter.

SQ, or spiritual intelligence, underpins IQ and EQ. Spiritual intelligence is an ability to access higher meanings, values, abiding purposes, and unconscious aspects of the self and to embed these meanings, values, and purposes in living richer and more creative lives. Signs of high SQ include an ability to think out of the box, humility, and an access to energies that come from something beyond the ego, beyond just me and my day-to-day concerns.

1.3   Three kinds of Thinking, Three kinds of Intelligences

Three kinds of neural organization in the brain which allow three different kinds of thinking, corresponding to the three kinds of intelligences.

We now know we grow neural connections throughout our lives; not the case that we start off with a fixed number and then lose them. Hence growth in children's intellectual capacities.

We think in different ways with different parts of the brain. We think with our heads but also with our emotions/bodies, and also with our spirits. Brain contains 10-100,000,000,000 neurons; there are 100 different sorts of cells. Each cell has roots, cell body, trunk, branches. Sensory inputs arrive at the roots, reach cell body if strong enough, fire along axon like a lit fuse, reach terminals, and jump over synapses to neighboring neurons.

Synapses work by chemical signaling; and use over a dozen different chemicals to do the job.

(i) Serial Thinking: IQ

Done along neural tracts – like phone cables, or chain of Christmas tree lights wired serially. Serial processing is what we use for rational thought processes, and this is what IQ tests measure. Examples are mental arithmetic, strategic planning.

(ii) Associative Thinking: EQ

This kind of thinking helps us form associations between things like hunger and food, mother and love, dogs and danger. It underlies EQ. It enables us to recognize patterns like faces, smells; learn skills like bike riding, piano playing. It is thinking with the heart/body. It is done in neural networks. Each contains bundles of up to 100,000 neurons, and each neuron may be connected to as many as 1000 others. Neural networks have ability to rewire selves with experience. Each time I see a pattern, the network connections which recognize it grow stronger. Ones that fire together become more strongly interconnected – for example, as we learn to drive a car (touch type?!). Associative learning is done by trial and error. Most emotions are trial and error; once I learn to feel angry at a certain stimulus, it is hard to react differently next time. Psychotherapy helps people break longstanding but inappropriate emotional association. Associative intelligence, including emotions, are not immediately verbal; hard to talk about them. We have 2 memory systems, one is based on precise neural wiring in the hippocampus, the other is based on associative neural networks located throughout the brain. First one subject to decay with age; second not. So hard to teach an older person serially wired skills, but not motor skills. Recent memory is in the first; long term memory, including emotions, based in the second. Emotional reactions have an associative base.

Associative thinking learns as it goes; but it tends to be habit bound. Hard to relearn an emotional response. And hard to share associative thinking with others.

Co-operation between IQ and EQ: ordinary chess players use IQ, serial thinking, only. Grandmasters use IQ and EQ – associative thinking, pattern recognition, too.

(iii) Unitive Thinking: SQ

Computers can do both serial and associative thinking. But they aren't conscious. We have a third kind of thinking, which is creative, insightful, intuitive. We learn and understand with IQ and EQ, but we invent and create with SQ. Neuroscientists have long been occupied with what they call the 'binding problem' – how do we put everything we take in through tracts and networks into a single coherent whole? Doctors have now isolated synchronous oscillations which pass over specific areas of the brain, that is, electrical signals oscillating at various frequencies. ECGs of people meditating show coherent brain waves across large areas of the brain. Magneto-encephalography is a new technology which has enabled detection of oscillations at 40 Hz over the whole brain. It is postulated that these enable information processing between the serial and parallel neural systems in the brain; provide a neural basis for consciousness itself; and are the neural basis for SQ.

1.4   Significance of Spiritual Intelligence

Spiritual intelligence is a path that leads us from fear and defensiveness to love and compassion; from ignorance and confusion to wisdom and understanding. A person equipped with 'spiritual intelligence' displays greater sense of commitment, integrity and empathy towards people or environment around him/her.

Spiritual intelligence is an attempt to bring into limelight the **fusion of four quotients** in an individual namely: (PQ) physical intelligence (body); (IQ) intelligence quotient (cognitive reasoning); (EQ) emotional Intelligence (emotional feelings) and (SQ) spiritual intelligence (transcendence). Spiritual intelligence helps us understand the mind and soul. It is being consciously or unconsciously practiced by almost all of us because it is the intelligence we use to determine what is right or wrong. It enables an individual to think dispassionately through issues from other people's perspective, and come to terms through the use of emotional bond between them.

1.5   Origins of Spiritual Intelligence

Danah Zohar coined the term "spiritual intelligence" and introduced the idea in 1997 in her book 'Rewiring the Corporate Brain'.

In the same year, 1997, **Ken O'Donnell**, an Australian author and consultant living in Brazil, also introduced the term "spiritual intelligence" in his book '**Endo quality**' - the emotional and spiritual dimensions of the human being in organizations.

In 2000, in the book, 'Spiritual Intelligence', author **Steven Benedict** outlined the concept as a perspective offering a way to bring together the spiritual and the material, that is ultimately concerned with the well-being of the universe and all who live there.

Howard Gardner, the originator of the theory of multiple intelligences, chose not to include spiritual intelligence in his "intelligences" due to the challenge of codifying quantifiable scientific criteria. Instead, Gardner suggested an "existential intelligence" as viable. The contemporary researchers continue to explore the viability of Spiritual Intelligence (often abbreviated as "SQ" or "SI") and to create tools for measuring and developing it. So far, the measurement of spiritual intelligence has tended to rely on **self-assessment instruments**, which can be susceptible to false or unreliable reporting.

However, in his 2009 doctoral dissertation, **Yosi Amram** found that the self-reported measure of spiritual intelligence predicted leadership effectiveness as rated by outside observers. In this research, he also deployed 360-assessments of spiritual intelligence and emotional intelligence, finding the observer ratings of SI to predict the leadership effectiveness ratings from other observers, offering predictive validity for SI even when controlling for emotional intelligence. Studies by other researchers have shown that leaders' SI can predict a variety of positive outcomes, such as financial performance of their organizations. Such cross-method studies lend overall validity to the construct of spiritual intelligence and its self- and 360-assessments.

A broad review of the research on SI has shown that:

(i)   several valid measurement instruments exist,

(ii)   they offer positive incremental predictive validity across a variety of desirable outcomes, and

(iii)   there is a neurological and biological basis for Spiritual Intelligence, highlighting the plausibility of its evolutionary adaptability, all of which supports SI's validity as an intelligence.

1.6   Definitions of Spiritual Intelligence

Definitions of spiritual intelligence rely on the concept of spirituality as being distinct from religiosity - existential intelligence.

1.6.1   Danah Zohar defined 12 principles underlying spiritual intelligence:

(i)   **Self-awareness:** Knowing what I believe in and value, and what deeply motivates me.

(ii)   **Spontaneity:** Living in and being responsive to the moment.

(iii)   **Being vision- and value-led:** Acting from principles and deep beliefs, and living accordingly.

(iv)   **Holism:** Seeing larger patterns, relationships, and connections; having a sense of belonging.

(v)   **Compassion:** Having the quality of "feeling-with" and deep empathy.

(vi)   **Celebration of diversity:** Valuing other people for their differences, not despite them.

(vii)   **Field independence:** Standing against the crowd and having one's own convictions.

(viii) **Humility:** Having the sense of being a player in a larger drama, of one's true place in the world.

(ix) **Tendency to ask fundamental "Why?" questions:** Needing to understand things and get to the bottom of them.

(x) **Ability to reframe:** Standing back from a situation or problem and seeing the bigger picture or wider context.

(xi) **Positive use of adversity:** Learning and growing from mistakes, setbacks, and suffering.

(xii) **Sense of vocation**: Feeling called upon to serve, to give something back.

1.6.2 Ken O'Donnell, advocates the integration of spiritual intelligence (SQ) with both rational intelligence (IQ) and emotional intelligence (EQ). IQ helps us to interact with numbers, formulas and things, EQ helps us to interact with people and SQ helps us to maintain inner balance. To calculate one's level of SQ he suggests the following criteria:

(i) How much time, money and energy and thoughts do we need to obtain a desired result.

(ii) How much bilateral respect there exists in our relationships.

(iii) How "clean" a game we play with others.

(iv) How much dignity we retain in respecting the dignity of others.

(v) How tranquil we remain in spite of the workload.

(vi) How sensible our decisions are.

(vii) How stable we remain in the disturbing situations.

(viii) How easily we see virtues in others instead of defects.

1.6.3 One More Definition

Another author defines spiritual intelligence as "the adaptive use of spiritual information to facilitate everyday problem solving and goal attainment." He further argued that "spiritual intelligence can be viewed as a form of intelligence because it predicts functioning and adaptation and offers capabilities that enable people to solve problems and attain goals."

He proposed five components of spiritual intelligence:

(i)   The capacity to transcend the physical and material.

(ii)   The ability to experience heightened states of consciousness.

(iii)   The ability to sanctify everyday experience.

(iv)   The ability to utilize spiritual resources to solve problems.

(v)   The capacity to be virtuous.

The fifth capacity was later removed due to its focus on human behavior rather than ability, thereby not meeting previously established scientific criteria for intelligence.

1.6.4   Another Definition

Yet another author defined spiritual intelligence as "the human capacity to ask ultimate questions about the meaning of life, and to simultaneously experience the seamless connection between each of us and the world in which we live."

1.6.5   Yet Another Definition

According to yet another definition, "Spiritual intelligence is concerned with the inner life of mind and spirit and its relationship to being in the world."

Spiritual intelligence implies a capacity for a deep understanding of existential questions and insights into multiple levels of consciousness. Spiritual intelligence also implies awareness of spirit as the ground of being or as the creative life force of evolution. If the evolution of life from stardust to mineral, vegetable, animal, and human existence implies some form of intelligence rather than being a purely random process, it might be called spiritual. Spiritual intelligence emerges as the consciousness evolves into an ever-deepening awareness of matter, life, body, mind, soul, and spirit. Spiritual intelligence, then, is more than individual mental ability. It appears to connect the personal to the transpersonal and the self to spirit. Spiritual intelligence goes beyond conventional psychological development. In addition to self-awareness, it implies awareness of our relationship to the transcendent, to each other, to the earth and all beings. It may be expressed in and culture as love, wisdom, and service

Another author defined spiritual intelligence as the "ability to draw on one's spiritual abilities and resources to better identify, find meaning in, and resolve existential, spiritual, and practical issues. Such resources and abilities, be it prayer, intuition, or transcendence, ought to be relevant to facilitating an individual's capacity for finding meaning in experiences, for facilitating problem solving, and for enhancing an individual's capacity for adaptive decision making."

According to one **Indian author**: "Spiritual Intelligence," is about the growth of the human beings. It is about moving on in life. About having a direction in life and being able to heal ourselves of all the resentment we carry, it is thinking of ourselves as an expression of a higher reality. It is also about how we look at the resources available to us. We realize that the nature is not meant to be exploited. Ultimately, we discover freedom from our sense of limitation as human being and attain Moksha (salvation)".

1.6.6 Definition of 'Spiritual Intelligence' given by Yosi Amram

Yosi Amram defines spiritual intelligence as "the ability to apply and embody spiritual resources and qualities to enhance daily functioning and wellbeing." Based on interviews with seventy-one spiritual leaders nominated by their peers, his ecumenical grounded theory of spiritual intelligence as presented at the 115th Annual Conference of the American Psychological Association, highlights **seven major themes** that are universal across the world's spiritual and wisdom traditions. They are:

(i) **Consciousness:** Possessing developed, refined awareness and self-knowledge.

(ii) **Grace:** Living in alignment with the sacred, manifesting love for and trust in life.

(iii) **Meaning:** Experiencing significance in daily activities through a sense of purpose and a call for service, including in the face of pain and suffering.

(iv) **Transcendence:** Identifying beyond the separate egoic self into an interconnected wholeness.

(v) **Truth:** Living in open acceptance, curiosity, and love for all creation (all that is).

(vi) **Serenity:** Surrendering peacefully to Self (Truth, God, Absolute, true nature).

(vii) **Inner-Directedness:** Maintaining inner-freedom aligned with responsible, wise action.

1.6.7 Research on 'Spiritual Intelligence' by David King

David B. King did research on spiritual intelligence at Trent University in Peterborough, Ontario, Canada. He defines spiritual intelligence as a set of adaptive mental capacities based on non-material and transcendent aspects of reality, specifically those that:

> "… contribute to the awareness, integration, and adaptive application of the nonmaterial and transcendent aspects of one's existence, leading to such outcomes as deep existential reflection, enhancement of meaning, recognition of a transcendent self, and mastery of spiritual states."

King further proposes four core abilities or capacities of spiritual intelligence:

(i) **Critical Existential Thinking:** The capacity to critically contemplate the nature of existence, reality, the universe, space, time, and other existential/metaphysical issues; also the capacity to contemplate non-existential issues in relation to one's existence (i.e., from an existential perspective).

(ii) **Personal Meaning Production:** The ability to derive personal meaning and purpose from all physical and mental experiences, including the capacity to create and master a life purpose.

(iii) **Transcendental Awareness:** The capacity to identify transcendent dimensions/patterns of the self (i.e., a transpersonal or transcendent self), of others, and of the physical world (e.g., nonmaterialism) during normal states of consciousness, accompanied by the capacity to identify their relationship to one's self and to the physical.

(iv) **Conscious State Expansion:** The ability to enter and exit higher states of consciousness (e.g. pure consciousness, cosmic consciousness, unity, oneness) and other states of trance at one's own discretion (as in deep contemplation, meditation, prayer, etc.).

1.6.8 Research on 'Spiritual Intelligence' by Two Indian Authors

Two Indian authors have also researched this concept extensively. Operationalizing the construct, they defined spiritual intelligence as "the capacity of an individual to possess a socially relevant purpose in life by understanding 'self' and having a high degree of conscience, compassion and commitment to human values."

1.7   Twenty-one Skills of Spiritual Intelligence

According to **Wigglesworth**, there are the following 21 skills of spiritual intelligence:

1.7.1   Quadrant 1: Self/self-Awareness

(i) Awareness of own worldview:

Do you feel that you can explain to others the impacts of your culture, your upbringing, and your mental assumptions on how you interpret the world around you?

(ii) Awareness of Life Purpose:

Do you feel that you can explain your life purpose to others? Do you stay focused on it consistently?

(iii) Awareness of Values Hierarchy:

Can you name and rank your top 5 personal values? Do you keep them in mind when making important choices?

(iv) Complexity of Inner Thought:

Can you hold conflicting perspectives on the "right thing to do" simultaneously? Can you make decisions in the face of uncertainty?

(v) Awareness of Ego self/Higher Self:

Can you consistently hear the voice of your Higher Self?

1.7.2   Quadrant 2: Universal Awareness

(vi) Awareness of Interconnectedness of Life:

Do you feel the pain of (do you deeply empathize with) other humans and animals who are suffering? Do you consider the consequences of your choices on ecosystems and future generations?

(vii) Awareness of Worldviews of others:

Do you seek to understand the emotions and perspectives of others even if you disagree with them? Do others feel understood by you?

(viii) Breadth of Time Perception:

Do you consider the history that brought you to the worldview you have today? Can you hold a billion years of history in your mind and perceive an evolutionary trajectory in the universe?

(ix) Awareness of Limitations/Power of Human Perception:

Are you aware of how your senses give you incomplete and sometimes inaccurate information? Do you supplement your five senses with intuition or spiritual insight?

(x) Awareness of Spiritual Laws:

Do you think about and experiment with spiritual laws/principles? So, do you try to live by your understanding of spiritual laws?

(xi) Experience of Transcendent Oneness:

Have you ever experienced a moment of awe, wonder, or non-ordinary consciousness? Has this experience of something transcendent helped you to focus on living from your Higher Self?

1.7.3   Quadrant 3: Self/Self Mastery

(xii) Commitment to Spiritual Growth:

I am willing to learn about spiritual topics from many sources. I commit time and energy to my own spiritual growth.

(xiii) Keeping Higher Self in Charge:

I am able to shift intentionally from listening to the voice of my ego to listening to my Higher Self. My Higher Self voice is clear and is the primary voice I hear.

(xiv) Living Your Purpose and Values:

My purpose and values are aligned with my Higher Self. My actions, decisions, and goals are aligned with my higher purpose and values.

(xv) Sustaining Faith:

I trust that there is a wise and loving nature to Life/ the universe/ all that is. I maintain an attitude of gratitude even when faced with difficulties.

(xvi) Seeking Guidance from Higher Self:

I actively seek guidance from sources beyond my own logic or ego. This includes seeking the wisdom of people I respect, of great teachers/writings, and from my Higher Self or Higher Power.

1.7.4   Quadrant 4: Social Master/Spiritual Presence

(xvii) Being a Wise and Effective Teacher/ Mentor of Spiritual Principles:

I enjoy teaching about spiritual principles. I do that through walking my talk and awakening the learner in other people.

(xviii) Being a Wise and Effective Leader/ Change Agent:

I can see and feel the perspectives of all the parties involved in a change. I am able to release my need to control or have things my way.

(xix) Making Compassionate and Wise Decisions:

I am compassionate toward my own mistakes as well as those made by others. I know how to set boundaries when I need to do so.

(xx) Being a Calming, Healing, Presence:

Other people feel calmer in my presence.

(xxi) Being Aligned with the Ebb and Flow of Life:

I instinctively know what is trying to come into form, and I can apply the right amount of action when it is needed to assist the process.

1.8   Advantages of having Spiritual Intelligence

(i) Enhances Emotional Intelligence

Spiritual intelligence enables individuals to cultivate a deeper understanding of their own emotions and those of others. This understanding enables individuals to better regulate their emotions and respond to others in a more compassionate and empathetic manner. This leads to greater emotional stability, which is crucial in navigating the challenges and uncertainties of the global economy.

(ii) Develops Resilience

Spiritual intelligence helps individuals develop resilience, which is the ability to bounce back from setbacks and challenges. It provides a sense of purpose, meaning, and connection to something greater than oneself, which can sustain individuals during difficult times. This is particularly important in the global economy as well as interpersonal relationships where changes are rapid and constant, and success often depends on the ability to adapt to new circumstances.

(iii) Promotes Creativity and Innovation

Spiritual intelligence encourages individuals to be more open-minded and to explore new perspectives and ideas. This, in turn, can lead to greater creativity and innovation. In a rapidly changing global economy, this is essential for individuals and organizations to remain competitive and relevant.

(iv) Improves Decision Making

Spiritual intelligence helps individuals to think more deeply and critically about the decisions they make. It provides a framework for considering the ethical and moral implications of decisions, which can help individuals make more informed and responsible choices. In situations, where decisions can have far-reaching consequences, this is particularly important.

(v) Encourages Personal Growth and Development

Spiritual intelligence encourages individuals to continuously reflect on their own values and beliefs and to seek a deeper understanding of themselves and their place in the world. This leads to personal growth and development.

(vi) Fosters Positive Relationships

Spiritual intelligence helps individuals cultivate positive relationships with others based on mutual respect, empathy, and understanding. This leads to greater collaboration and teamwork, which is crucial harmonious living in the family and society.

1.9 Measurement of Spiritual Intelligence

1.9.1 Measurement of spiritual intelligence often relies on self-reporting. **Yosi Amram** and Christopher Dryer developed and validated the **Integrated Spiritual Intelligence Scale (ISIS)**—a self-report measure of spiritual intelligence (also applied as a **360-assessment measure**), which showed satisfactory factor structure, internal consistency, test-retest reliability, and construct validity. Applications of the ISIS by other researchers has shown that "the reliability of ISIS was high (i.e., Cronbach's alpha = 0.97)" and that there was "a significant positive relationship between employees' spiritual intelligence and work satisfaction." It consists of twenty-two sub-scales assessing separate SI capabilities related to Beauty, Discernment, Ego-lessness, Equanimity, Freedom, Gratitude, Higher-self, Holism, Immanence, Inner-wholeness, Intuition, Joy, Mindfulness, Openness, Practice, Presence, Purpose, Relatedness, Sacredness, Service, Synthesis, and Trust. These twenty-two sub-scales are grouped into five domains: Consciousness, Grace, Meaning, Transcendence and Truth. ISIS has since been translated into several other languages and validated by other researchers.

1.9.2 Two researchers have developed a self-report measure, **the Spiritual Intelligence Self-Report Inventory (SISRI-24)** with psychometric and statistical support across two large university samples. Yet another researcher has developed the SQ21, a self-assessment inventory that has tested positively for criterion validity and construct validity in statistically significant samples. Wigglesworth's SQ model and assessment instrument have been successfully used in corporate settings.

1.9.3 The **Scale for Spiritual Intelligence (SSI)**, developed by two Indian authors, is a 20-item, self-report measure of spiritual intelligence in adolescents. The idea behind the development of this scale was to generate and assess the concept of spiritual intelligence in the collectivist culture bounded with eastern philosophy. The SSI is rated on a Likert scale and can be completed in 10 minutes.

1.9.4 The 29-item Spiritual Intelligence Questionnaire

This test was normalized by three researchers on students. The normal group was 280 people, 200 of whom were the students of Gorgan University of Natural Resources and 80 students of Payame Noor University of Behshahr. Of these, 184 were female and 96 were male. First, a 30-item questionnaire was prepared by the test developers and implemented on 30 students. The reliability of the test in the initial phase was 0.87 by the alpha method. In the analysis of the question by Loop method, 12 questions were removed and the final questionnaire was adjusted with 29 phrases. At the final stage, the questionnaire was implemented on 280 subjects and the reliability was 0.89 at this stage. Factor analysis was used to evaluate validity in addition to formal content validity that the questions were confirmed by the experts (colleagues) and the correlation of all questions was higher than 0.3. In Varimax rotation, two major factors were found to reduce variables. The first factor with 12 questions was called "understanding and communicating with the source of universe" and the second factor with 17 items was called "spiritual life or reliance on the inner core."

1.10 Components of Spiritual Intelligence

(i)   Knowing and believing in God divinity.

(ii)   Making work and life meaningful (for the sake of Almighty/ Creator)

(iii)   Deep self-awareness.

(iv)   Spiritual wisdom.

(v)   Having lofty goals in life and work.

(vi)   The ability to use the intellectual resources to solve problems in life.

(vii)   The capacity for virtuous behavior (forgiveness, charity, humility, gratitude, etc.).

(viii) Willingness to serve.

(ix)   The ability to align individual, organizational, and societal goals.

(x)   Having insight, understanding, diagnosing, and distinguishing.

(xi)   Work conscience.

(xii)   Systematic thinking (regarding to dimensions of human spirit).

(xiii)   Paying attention to spiritual values.

(xiv)   Having positive feelings.

(xv)   The ability to ask fundamental questions and find fundamental answers.

One of the aspects of spiritual intelligence is understanding and insight of the people about themselves by which they regulate their own emotions. On the other hand, without having a high level of emotional intelligence, the people cannot possess the attributes such as honesty, forgiveness, etc. Other components of spiritual intelligence include patience, spiritual and religious beliefs and practices, meaning and having purpose in life, divinity, inner calm, spiritual experiences, self-scrutiny, as the main components of spiritual intelligence. Spiritual intelligence is the human capacity for asking questions about the meaning of life and simultaneously experiencing the connection between each of us and the world in which we live.

1.10.1   Components of Mental Toughness

There is a two-way relationship between mental toughness and the components of spiritual intelligence.

Following major components of mental toughness, which have been identified, are: (i) self-confidence, (ii) faith in oneself abilities, (iii) faith in one self's abilities to realize his goals, (iv) the desire for success, (v) the ability to recover after defeats, (vi) the ability to endure stress and physical pain, (vii) the ability to control worry, flourishing, (viii) the ability to have self-control after unexpected events, and (ix) the intense concentration. These authors believed that mental toughness consists of four cornerstones: faith in oneself, controlling pressures, concentration, and motivation. Further, it was indicated that mental toughness consists of nine components: faith, adaptation, concentration, self-motivation, self-control, intelligence, flexibility, personal values, and physical toughness. It indicates mental toughness includes five components: self-confidence, self-control, self-discipline, worry control, and concentration. The studies state that mental toughness is positively related to many positive variables like stress evaluation, logical thinking, problem-solving strategies, self-confidence, self-control, and predictor of academic success. Mental toughness is also related the social solidarity, positive attitude, and a meaningful life.

1.11   Seven Factors of Spiritual intelligence

(i)   **Caring:** paying attention to physical processes such as eating, regular meditation, and exercises such as yoga and tai-chi.

(ii)   **Enlightenment:** involving of mind in reading spiritual issue, sacred texts and analyzing them.

(iii)   **Divinity:** the sense of connection with God, a higher power, and with a source of Divine power.

(iv)   **Spirituality in childhood:** the intellectual interests and activities in childhood such as doing spiritual practices ceremony and reading sacred texts.

(v)   **Ultra-sensory perception:** the experiences that are referred to supernatural or sixth sense.

(vi)   **Psychological trauma:** spiritual awareness which are reached by having painful experiences (optional).

(vii) Paying attention to community: performing social activities such as donation, or volunteering activities which will be beneficial to the community.

1.12   Six Aspects of Spiritual Intelligence

Spiritual intelligence is defined as the ability of the individual to understand existential issues and to employ their spiritual potentialities in facing real-life challenges, which scaffold their self-confidence.

It includes the following aspects:

(i) **Awareness:** the ability of the individual to know himself, his beliefs, spiritual laws, and the causes of his existence.

(ii) **Contemplation:** the ability of the individual to think critically of existence and deduce the evidence of the supernatural power of the creator and His greatness.

(iii) **Positive diagnosis of the difficulties:** the ability of the individual to think of the obstacles that he faces as success opportunities and to have a positive vision toward life.

(iv) **Transcendence:** the ability of an individual to transcend selfish interests and to look for a set of transcendental values establishing happiness for him and others.

(v) **Spiritual practices:** the ability of an individual to practice the sacred rituals imposed upon him as a kind of gratitude to God.

(vi) **Good morals:** the individual should have lots of good virtues like mercy, tolerance, and modesty that enable him to make charitable deeds.

1.13   Seven Steps to Spiritual Intelligence

R. Bowell, in his book, "Seven Steps to Spiritual Intelligence" explains that spiritual intelligence seeks to discover the why of what we do, rather than what we do or how we do it. He explores the following seven steps to spiritual intelligence:

(i) Awareness

The first step towards spiritual Intelligence is awareness. Sense organs play an important role in bringing about awareness. It promotes recognition, associations, memories, dialogues etc., The main aim of this step is to make one individual to become aware about what he has not yet seen or heard or noticed. In this step, spiritually intelligent person will refrain from the relaxed state and keep oneself awaken to great challenges and adventures of evolving life.

(ii) Meaning

This step emphasizes on value judgments in one's life. It develops consciousness of the world and all that lives within it. The meaning of things cannot be assumed.

(iii) Evaluation

Evaluation should never be done from the identity level of 'self' alone. It is a process by which one understands her/his self and this is to understand the other person too.

(iv) Being Centred

Being centered is to occupy a higher level of engagement altogether. This step makes an individual committed to the growth of 'self' as a meaningful life.

(v) Vision

Seeing what others have not yet seen is a sign of visionary. One should have a vision that can see beyond the materialistic world. This step helps in developing consciousness about the situation.

(vi) Projection

Action is followed by projection. Projection begins in the settlement of 'self', and in the vision of great wealth that can be achieved, when one truly sees the truth of the situation.

(vii) Mission

This final step towards spiritual intelligence integrates one's self with the truth of the situation. Mission statement is an important aspect of the corporate identity and it inspires those who follow.

1.14   Eleven Dimensions of Spiritual Intelligence

(i)   Internal knowledge.

(ii)   Deep intuition.

(iii)   Self-awareness and integrating with nature and the universe.

(iv)   Ability to solving problems dispassionately.

(v)   Empathy towards others.

(vi)   Inner guidance and using sublimation methods such as intuition to solve problems.

(vii)   Holistic view in order to see the interconnection among different aspects.

(viii) Acceptance and love of truth, living in synchronization with the creator/nature.

(ix)   Spiritual Surrender to Almighty/Super power/Nature.

(x)   Spiritual Maturity.

(xi)   Spiritual Flexibility.

1.15   More on Danah Zohar's 12 principles of Spiritual Intelligence

(i) Self-Awareness

Spiritual self-awareness means to recognize what I care about, what I live for, and what I would die for. It's to live true to myself while respecting others. Being authentic in this way is the bedrock of genuine communication with our deeper self that allows us to bring that self into the outer world of action.

(ii) Spontaneity

Being spontaneous does not mean merely acting on a whim but refers to behavior honed by the self-discipline, practice, and self-control of the martial arts warrior. To be spontaneous means letting go of all your baggage — your childhood problems, prejudices, assumptions, values, and projections — and be responsive to the moment. And since spontaneity comes from the same Latin root as responsibility, it means taking responsibility for our actions in the moment.

(iii) Being Vision and Value-Led

Vision is the capacity to see something that inspires us and means something broader than a company vision or a vision for educational development. It seeks answers to the bigger, more difficult questions such as Why do we want the world to have our products? and What are we trying to educate children for?

Follow what you want to do. But whatever you do, make a difference with it." That's having a life that's led by vision and values.

(iv) Holism

In quantum physics, holism refers to systems that are so integrated that each part is defined by every other part of the system. As I stand here in this room, which is a system, the words that I say, the tone of my voice is partly brought out by speaking to you. And you are partly responding to me. For the moment that we are together this morning, we are defined in terms of each other. What I think, feel, and value affects the whole world. Holism encourages cooperation, because as you realize you're all part of the same system, you take responsibility for your part in it. A lack of holism encourages competition, which encourages separateness. For more effective collaborations, we need cooperation and a sense of oneness.

(v) Compassion

In Latin, compassion is defined as "feeling with." I don't just recognize or accept your feelings, I feel them. This is particularly hard to do with someone who has hurt you. Can you feel the pain and frustration behind their behavior? You don't have to let them treat you that way, and often you do have to fight. But fight with compassion, with understanding, with knowledge of your enemy.

(vi) Celebration of Diversity

Compassion is strongly linked to the principle of diversity. We celebrate our differences because they teach us what matters.

When someone disagrees with you, he or she literally makes you grow new neurons. You have to rewire my brain, challenge your assumptions, and question your values. When a group experiences divisive, painful issues, some people ask, Dare we confront them? Mightn't they split us? Shouldn't we put aside our differences and see what we can agree about? Absolutely not. Celebrate the differences. Cauterize the pain by letting it come out. That's where the passion and energy is in our collaborations. You'll find that, if you do it in a dialogic spirit, the collaboration becomes a container that can hold all that diversity and allow it to emerge into something new. By not bringing it into the group, you lose that energy. Celebrating diversity means that I appreciate that you rattle my cage, because by doing so, you make me think and grow.

(vii) Field Independence

Field independence is a term from psychology that means "to stand against the crowd," to be willing to be unpopular for what I believe in. It's a willingness to go it alone, but only after I've carefully considered what others have to say.

(viii) Humility

Humility is the necessary other side of field independence, whereby I realize that I am one actor in a larger play and that I might be wrong. So I question myself ruthlessly. Am I right to think what I do? Have I listened to all the arguments against it? Have I thought deeply about it? Humility makes us great, not small. It makes us proud to be a voice in a choir.

(ix) Tendency to Ask Fundamental "Why?" Questions

"Why?" is subversive, and people are often frightened by questions without easy answers. Why are we doing it this way rather than that way? Why am I in this collaboration, and what does it exist for? Why aren't we doing something else? Einstein said that as a boy he was in trouble all the time at school because the teachers accused him of asking stupid questions. When he became famous, he joked that now that everybody thought he was a genius, he was allowed to ask all the stupid questions he liked. Answers are a finite game; they're played within boundaries, rules, and expectations. Questions are an infinite game; they play with the boundaries, they define them.

(x) Ability to Reframe

Reframing refers to the ability to stand back from a situation and look for the bigger picture. One of the greatest problems of our world today is short-term thinking. As those of you from the business community know, most corporations keep an eye on three months down the road when the quarterly returns come in and shareholder value is paid out.

(xi) Positive use of Adversity

This principle is about owning, recognizing, accepting, and acknowledging mistakes. How many of us get trapped in courses of action because the initial step we took was a mistake and we didn't want to lose face by admitting it? Rather than having the courage to acknowledge our error, we pursue the mistaken course of action, digging ourselves deeper into the mess. Have you ever admitted a mistake to someone where it really hurt to do so? Have you felt the energy flow out of you when you admitted it? Great passion and energy can be released by saying the simple words "I made a mistake. What I did was wrong, and therefore I'm now going to embark on a different course."

Positive use of adversity is also the ability to recognize that suffering is inevitable in life. There are painful things for human beings to deal with, yet they make us stronger, wiser, and braver. How boring we would be if we never had any adversity in our lives!

(xii) Sense of Vocation

This principle sums up spiritual intelligence and spiritual capital. Vocation comes from the Latin 'vocare', "to be called." Originally, it referred to a priest's calling to God. Today it often refers to the professions such as medicine, teaching, and law. The business will become a vocation that appeals to people with a larger purpose and a desire to make wealth that benefits not only those who create it but also the community and the world.

The expected outcomes of a person having a high 'spiritual intelligence are: happiness, serenity, good self-esteem and harmonious and loving relationships, ability to think out of box, modesty, and an access to energies beyond ego and beyond day-to-day concerns. Spiritual intelligence is the way we assign meaning and feel connected to the higher power creator or God,. Spiritual intelligence is a set of adaptive mental capacities based on non-material and transcendent aspects of reality.

1.16   Development of Spiritual Intelligence

Some researchers approach to the intelligences, like identification of motive- and trait-level competencies that predict superior performance, seems to refer mainly to the strength of innate traits differentially distributed across the population. However, even traits can be developed, just as native cognitive intelligence (IQ), for example, can be honed by training in logic, strategy, analogies, and other forms of reasoning and problem-solving. So, too, can Spiritual Intelligence, at least as assessed by some of the instruments.

For example, it was found that SI was significantly correlated with age in their original survey samples, so that the more mature a person, the greater the likelihood of higher SI. Very few studies have examined this relationship over a widely age-varied population. Even less research has been conducted on the relationship between general education and SI, but one researcher (2012) found that parents with higher education scored higher on SI than those with primary and secondary education.

Specific training in SI constructs produces measurable differences. In healthcare, some researchers, most of them in Iran, have tried to develop SI as way to help people cope with existential challenges. For example, two independent studies showed that training in Spiritual Intelligence and subsequent assessment have a positive effect on hope and life expectancy among the chronically ill. Others have trained care providers in SI to work with patients. It was found that SI training improved nurses' competence in spiritual care in critical care units, which in turn improved patient quality of life. And SI training interventions have been shown to improve the quality of work life among healthcare providers. A metanalysis of seven studies comprising 512 nurses and nursing students showed that those who received SI training demonstrated higher SI scores at two- and four-week follow-up compared to controls, and that the experimental group scored significantly higher in communication skills, job satisfaction, and spiritual care competence and significantly lower in overall stress compared to controls. Significantly higher job satisfaction was reported at two-month follow-up by those who had had SI training.

Research conducted outside a healthcare setting examined whether mental health generally could be improved through SI training. In a study (2014) of the effect of SI training on the mental health of Iranian high school students, showed that the intervention decreased interpersonal sensitivity, somatization, obsessive-compulsive behavior, depression, anxiety, aggression, phobia, paranoid ideation, and psychoticism in the experimental group compared with controls.

Thus, it appears that like other intelligences, innate SI capacities, which theoretically should be differentially distributed in a population like variations in IQ, can nevertheless be cultivated to some extent, especially with focused training.

We can develop our spiritual intelligence through study and practice of skills like:

(i)   Quieting the mind,

(ii)   Tapping into our inner wisdom and intuition,

(iii)   Cultivating gratitude,

(iv)   Practicing humility and empathy,

(v)   Connecting intimately with God and others.

1.17   Biological and Genetical Bases for Spiritual Intelligence

Spiritual neuroscience is a new area of study that examines religious and spiritual experiences and behavior as correlates of brain function, including alleging that this connection indicates an evolutionary basis for spirituality—i.e., that spirituality serves human adaptation. Indeed, some spiritual qualities appear to have genetic roots. Reducing spirituality to biological structures is controversial, but the growing body of research on the effects of spiritual practices on brain functioning and development indicates that such correlations do exist.

Rather than a single "God spot" in the brain, several neural networks, structures, and genetic components seem to be involved. One researcher (2020) concluded that religious or spiritual cognition involves a complex interplay among specific regions in the brain. Another researcher (2022) found that spirituality maps to a neural brain circuit in the periaqueductal gray region.

A meta-analysis of 80 studies suggested that empathy is mediated using six spatially distinct activation clusters in the medial part of the frontal lobe dorsal to the inter-commissural plane. Empathy and compassion cultivated through particular spiritual practices change brain activity and structure. Some researchers (2004) found that a long-term compassion and loving-kindness meditation practice is associated with altered resting electroencephalogram patterns, suggesting that the development of SI relatedness qualities involves temporal integrative mechanisms and may induce short- and long-term neurological changes. Furthermore, they found specific brain area activation during loving-kindness and compassion meditation among such trained meditators.

Similarly various parts of the brain have been associated with transcendent experiences using positron emission tomography

(PET) scans, found an inverse correlation between serotonin receptor density and experiences of self-transcendence. Two researchers (2021) argued that the neural underpinnings of transcendent thought capacities are present and developed as adolescents actively make meaning, rely on supportive social relationships, and process deeper reflections.

Genetics may also play a part in SI. One researcher (2004) found a gene contributing to the self- reported value of self-transcendence from his study of same-sex siblings. Another researcher (1999) found genetic factors to be important in influencing self-transcendence based on a study of Australian twins.

1.18   Healing Ourselves with Spiritual Intelligence

Zohar and Marshall described the harmony of self and spiritual intelligence using the petal layers of the lotus, which has long been a symbol of spiritual and physical integration in Eastern traditions. The lotus is to represent the three layers of self: (a) ego, (b) associative unconscious, and (c) inner self. The ego layer is tied to personality types based on Holland and Jung's personality type interrelationships. The associative unconscious, likened to Freud's id, is representative of the symbolic realm of the unconsciousness where one houses images, habits, cultural influences, and so forth. Inner self, the final lotus layer, is composed of one's dreams, imagination, and one's spirit. In their later work, Spiritual Capital, Zohar and Marshall tied SI to a value-based capitalistic culture with a scale of motivations based on Maslow's hierarchy. "Maslow later differentiated the growth need of self-actualization into self-actualized and self-transcendent. Spiritual illness is a condition of fragmentation, spiritual health one of wholeness. Spiritual Intelligence is the means by which we move from one to the other. Key activity is recollection. Children show high Spiritual Intelligence – always asking why. Any time we step outside our assumptions or habitual way of seeing things, any time we break through into some new insight that places our behavior in a larger, meaning-giving context, any time we transcend ego and act from our centre, any time we experience the thrill of beauty or truth larger than ourselves, hear the sublimity in a piece of music, see the majesty in mountain sun-rise, feel the profound simplicity of a new idea, feel the depths of meditation or the wonder of prayer, we are experiencing our Spiritual Intelligence and using it to heal ourselves. I think I am conversing with God when I do this.

People like **Mahatma Gandhi, Swami Ramakrishna Paramhansa, Swami Vivekananda, Sri Aurobindo** are some of those who followed their convictions courageously. It was the presence of not just their IQ and EQ but also SQ that rendered their thought process different from the others who strictly conformed to societal customs and practice.

(Late) Dr. A.P.J. Abdul Kalam, former President of India, asserted that "Our culture teaches us to learn both Para (spiritual) and Apara (Worldly) vidya (knowledge). Therefore, together with knowledge of the Apara, one should learn the para as well. If one learns this then Apara-(worldly knowledge-vidya) will become founded on dharma and spirituality. One must remember that in God's scheme of things, the whole purpose behind creation is the idea that every person, every soul attains bliss.

Spiritual Intelligence is the soul's Intelligence. It is the Intelligence with which we heal ourselves and with which we make ourselves whole. Spiritual Intelligence, in essence, represents a dynamic wholeness of self in which the self is at one with itself and with the whole of creation. SQ gives us our ability to discriminate. It gives us our moral sense, an ability to temper rigid rules with understanding and compassion and an equal ability to see when compassion and understanding have their limits.

We shall discuss in detail, many of the topics listed in this chapter, in the subsequent chapters of this book.

SECTION-II

Skills, Aspects, and Components of Spiritual Intelligence

CHAPTER 2

SELF-AWARENESS

Knowing yourself is life's eternal homework"

– Felicia Day

2.1   What is Self-Awareness?

Self-awareness means understanding one's thoughts, feelings, values and background and how they impact the success of the interaction and relationship, or how they may influence one's work. It is recognizing one's own biases by tracing them to their origins, through reflection and by noticing one's own behavior— and then intentionally seeking a way forward that positively impacts the interaction and relationship. It means exploring new ways of thinking and acting when situations become difficult or uncertain, or in times of urgency. Self-awareness is your ability to perceive and understand the things that make you who you are as an individual, including your personality, actions, values, beliefs, emotions, and thoughts. Essentially, it is a psychological state in which the self becomes the focus of attention.

While self-awareness is central to who you are, it is not something you are acutely focused on at every moment of every day. Instead, self-awareness becomes woven into the fabric of who you are and emerges at different points depending on the situation and your personality.

It is one of the first components of the self-concept to emerge. People are not born completely self-aware. Yet evidence suggests that infants do have a rudimentary sense of self-awareness.

Infants possess the awareness that they are separate beings from others, which is evidenced by behaviors such as the rooting reflex in which an infant searches for a nipple when something brushes against their face. Researchers have also found that even newborns are able to differentiate between self- and non-self-touch. Studies have demonstrated that a more complex sense of self-awareness emerges around one year of age and becomes much more developed by approximately 18 months of age. The researchers applied a red dot to an infant's nose and then held the child up to a mirror. Children who recognized themselves in the mirror would reach for their own noses rather than the reflection in the mirror, which indicated that they had at least some level of self-awareness.

Self-awareness is being self-aware is all about having an understanding of your own thoughts, feelings, values, beliefs, and actions. It means that you understand who you are, what you want, how you feel, and why you do the things that you do.

Researchers have proposed that an area of the brain known as the anterior cingulate cortex located in the frontal lobe region plays an important role in developing self-awareness. Studies have also used brain imaging to show that this region becomes activated in adults who are self-aware.

2.1.1   Self-Awareness demonstrates the behavior when:

- one observes own behavior and actions;
- one recognizes the impact of own behavior and actions upon others;
- one seeks feedback and considers it carefully;
- one is open to new perspectives and different ways of thinking and working;
- if uncomfortable with unfamiliar experiences, one expresses misgivings or discomfort and seeks help to move forward;

- one seeks guidance and support on ensuring personal perspective, and is sensitive and responsive to the needs and interests of one's people;

- one is aware of one's biases and monitors them to avoid misunderstanding;

- one manages one's thoughts and feelings when challenged. One looks at differences as opportunities, not threats;

- one speaks honestly about one's own biases and assumptions;

- one demonstrates awareness of personal biases when writing, or when interpreting the written word;

- one acknowledges the thinking, emotions and behaviors of others;

- one suspends judgment or decision making until fully understanding the situation;

- one shows emotions that are genuine, culturally appropriate and that honour the relationship;

- one continually looks for opportunities to improve self-awareness;

- one seeks self-awareness by spending time with own people, and elder people in one's community or in other settings;

- one recognizes how personal values arc shaped by one's ideas, belief systems and opinions;

- one recognizes assumptions and biases that surface when that value system is offended;

- one adapts to one's behavior in the moment to be more culturally appropriate and to honour the relationship;

- one avoids safe and predictable environments. One pushes self into uncomfortable and ambiguous situations to increase self-awareness;

- one notices and manages personal uncertainty and the fear of the unknown;

- one provides feedback that is truthful and candid. One coaches others towards self-awareness;

- one creates opportunities for the development of self-awareness, personally and in others.

Becoming more self-aware is the first step in aspects of personal growth.

Self-awareness takes humility and strength to allow oneself to open up, to dig deep within, and sit with what you see, feel, and observe. With awareness of self, you can regulate your interior condition, which has a ripple effect through your behavior and actions. It is a practice that can challenge your thoughts, your behavior and sense of self. When done consistently, you are better able to regulate your emotions and responses. From there, you are able to recognize where your thoughts and emotions are leading you, and hence, make the necessary changes you need. It helps in developing your spiritual intelligence.

2.2 Types of Self-Awareness

Psychologists often break self-awareness down into two different types, either public or private.

(i) Public Self-Awareness

This type emerges when people are aware of how they appear to others. Public self-awareness typically emerges in situations when people are at the centre of attention.

This type of self-awareness often compels people to adhere to social norms. When we are aware that we are being watched and evaluated, we often try to behave in ways that are socially acceptable and desirable.

Public self-awareness can also lead to evaluation anxiety in which people become distressed, anxious, or worried about how they are perceived by others.

You may experience public self-awareness in the workplace, when you're giving a big presentation. Or, you may experience it when telling a story to a group of friends.

Sometimes, people can become overly self-aware and veer into what is known as self-consciousness. Have you ever felt like everyone was watching you, judging your actions, and waiting to see what you will do next? This heightened state of self-awareness can leave you feeling awkward and nervous in some instances.

In a lot of cases, these feelings of self-consciousness are only temporary and arise in situations when we are "in the spotlight." For some people, however, excessive self-consciousness can reflect a chronic condition such as social anxiety disorder. While self-awareness plays a critical role in how we understand ourselves and how we relate to others and the world, excessive self-consciousness can result in challenges such as anxiety and stress.

(ii) Private Self-Awareness

This type of happens when people become aware of some aspects of themselves, but only in a private way. For example, seeing your face in the mirror is a type of private self-awareness.

Feeling your stomach lurch when you realize you forgot to study for an important test or feeling your heart flutter when you see someone you are attracted to are also examples of private self-awareness.

2.3   Elements of Self-Awareness

The five elements of self-awarenesses are:

(i) Consciousness

This means being aware of your internal experiences, including your emotions and thoughts.

(ii) Self-knowledge

This element is focused on your understanding of who you are, including your beliefs, values, and motivations.

(iii) Emotional intelligence

This element is focused on the ability to understand and manage emotions.

(iv) Self-acceptance

This aspect is centered on accepting who you are and showing yourself compassion and kindness.

(v) Self-reflection

This element of self-awareness involves being able to think deeply about your feelings, thoughts, and goals in order to gain an even better understanding of who you are and your place in the world.

2.4   How to Develop Self-Awareness?

- Be curious about who you are. How far you will go on your journey to understand yourself depends on what you are ready to explore and experience.

- Let your walls down. Try to let go of any judgement and the instinctual urge to protect yourself. Through a willingness and openness to yourself, you can let go of your defenses, thereby seeing yourself in different ways than what you have always assumed.

- Keep a journal and note what triggers positive feelings. This is also good practice of becoming mindful.

- Go-ahead, ask others how they see you. Be brave enough to receive feedback about yourself in various situations.

- If you are upset with someone, take a third person perspective. Often their experience will likely be different from yours. It is common for us to believe others will frame their situation in the same way as yours, but this is not often the case. We all come with various degrees of experiences and perspectives. Drop your defenses and ask others where they are at, be willing to receive feedback. This will also lend insight into awareness of yourself.

- Keep checking in with yourself. Clinically, the most effective method for self-awareness is to pause, and do a brief check in to where you are at in this moment. How are you feeling right now? What is driving this feeling?

- Remain open to continuous learning. Each experience will reveal things about yourself.

Ask yourself how is what I am learning serving me? What else do I need and what needs to happen for me to obtain this?

- Having self-awareness gives us the power to influence outcomes; helps us become better decision-makers and gives us more self-confidence. We can communicate with clarity and intention, which allows us to understand things from multiple perspectives. It frees us from assumptions and biases.

One can develop self-awareness through the following techniques:

(i) Meditation

Meditation can be an especially useful practice because you don't have to worry about changing anything—simply noticing what happens during a meditation can bring greater awareness of your thoughts and feelings.

Maybe you notice that you hold tension in your body by clenching your jaw, for instance, or that you tend to worry so much about the future that it's hard to be in the present moment. This is all valuable information that can help you get to know yourself and your tendencies.

(ii) Journaling

Journaling is a practice in self-reflection that can help you notice the ways in which you tend to think and behave, and even which areas in your life you may wish to improve. It can be a therapeutic way to gain insight into your life events and relationships.

(iii) Talk Therapy

During therapy—such as cognitive behavioral therapy (CBT)—a therapist works with you to address negative thought patterns or behaviors. By understanding the underlying cause of your negative thoughts, for instance, you're in a more advantageous position to change them and use healthy coping mechanisms instead.

2.5   Benefits of Developing Self-Awareness

- Being better able to manage and regulate your emotions;
- Better communication;
- Better decision-making skills;
- Improved relationships;

- Higher levels of happiness;
- More confidence;
- Better job satisfaction;
- Better leadership skills;
- Better overall perspective;
- More likely to make better choices.

2.6   What is Spiritual Awareness?

The Ancient Greek aphorism *"Know Thyself"* invites us to self-reflect; to look at ourselves and to understand our true nature, and having done so, to live our true nature. But what is our true nature? Are we simply the body that we come into this world with, that ages with time and withers away after death? Are we the mind that grows astute with intellectual pursuits only to evaporate at the time of death? Are we our emotions that change as frequently as the weather? Or is there much more to our "selves"? Is there something more constant, more powerful that defines our true self?

Saints and spiritual Masters, since time immemorial, have come to remind us that we are much more than the body that we see with our physical eyes, much more than our mind or emotions. They tell us that our true self is the spirit or soul. They tell us that the soul is a part of God, the Creator and that it is the Power that enlivens the body, that animates us, that makes us tick. When this Power leaves us, the body becomes motionless, and we are proclaimed dead. Saints also tell us that there is a way for each of us to experience our true nature. It is not an intellectual pursuit, but one of practice, wherein we can have a direct experience of our true self.

Currently, we are asleep and ignorant of this truth. To become spiritually aware is to awaken to our true nature as spirit, to experience our true self. Spirituality is the process of discovering our true self. We normally think that perception is possible only through our bodily sense organs. Nevertheless, when we become spiritually aware, we recognize that we can perceive with the spirit.

An empowered soul is a soul that has recognized itself and is aware that it is the essence of who we really are, that is the guiding power behind the body and mind.

We begin to understand that we are more than the body and mind- we are soul, filled with spiritual gifts far more valuable than any we can attain in this world. We become at peace because we learn the answers to the mysteries of life and death, and we know that we will survive beyond the demise of our physical body.

We experience love, peace, and eternal happiness. A whole new world opens up for us.

2.7   Self-Realization

Self-realization is a term used in Eastern religions, yoga philosophy, psychological theories and other spiritual schools of thought. It denotes a state in which an individual knows who they truly are and is fulfilled in that understanding.

For some, this may manifest itself in an understanding that they are at one with the omnipresence of God, or that the Divine is within them. Others consider it a fulfillment of all the possibilities of an individual's personality and character. The achievement of Self-realization may be seen as a scientific and/or spiritual process.

It is said that once a person reaches Self-realization, his new vision and understanding brings him continuous, permanent happiness. It will also bring:

- Equanimity to all circumstances,
- Inner peace,

- Freedom from all fears and anxieties,
- Deep spiritual fulfillment,
- Stronger, calmer relationships with others.

Self-realization cannot be achieved if you are struggling financially and too caught up in worrying about how to pay for the rent and provide food for your family. Unfortunately, this is usually the case for many people, which leaves little opportunity for them to maximize their abilities.

In religions, the concept of self-realization is taken from a different perspective altogether. Connecting with your truest self has a lot to do with transcending your own mind and body. This self is often considered as an eternal being that is not confined to the physical space that your mind and body take up. Many recognize this part of yourself as the soul.

To put all of these definitions together, self-realization is ultimately learning the answer to the foundational question, "Who am I?"

Once a person reaches Self-realization, he is freed from his own desires and worldly attachments. He is also liberated from external pressures, such as cultural and social expectations, or political and economic influences. He is beyond self-delusion and material attachments.

Self-realization involves letting go of many of the things that are associated with individual identity in order to find the true Self, which is eternal and unchanging. It is the difference between identifying with the ego and identifying with the true Self.

Who you really are is not even your body or your mind. These are all things you as a self-experience, but they are not you.

And when you are too caught up in these things that are not yours, that's when you fall victim to and get stuck in your negative experiences such as stress, anxiety and fear.

While your thoughts, feelings, and physical body always changes, you do not.

Self-realization is the truth of who we are, what we are — the realization that we are not the physical body, the physical form that we believe ourselves to be, but the energy within that physical form that gives us life. Most of us are unable to realize this because of the illusionary world we live in, because of our mind which makes us believe that we are the ego, and because of the senses that make us crave for sensual pleasures. Our mind limits us. It restricts our spiritual awareness. It is unable to focus and contemplate on the energy that lies beyond this physical form. For reasons unknown to us, we are ignorant of the truth. For instance, when we look at jewels in a jewelry shop, we see rings, bangles, chains, and we admire the jewelry. They seem to be rings, chains or bracelets but in reality, they are not, they are just gold. Gold appears as the ring, as the bangle and as the chain. But if you remove the gold, there is no jewelry. Realization that it is gold brings us to the truth that the jewelry is only an effect of the cause 'gold'.

2.7.1   What is that we realize?

The moment we are able to see this energy, which is beyond this physical form, we are realized. So, what is it that we realize? We realize that we are not the body, we are not the mind but we are indeed the life energy. The life energy is immortal, it never dies. The body dies and disintegrates, but the life energy, the Atman, the Chi, the Prana, Soul or the Spirit can merge with the Universal soul, the God on self-realization. Self-realization further leads to God-realization. God-realization is a journey of realizing the truth about God and thus, knowing God — the relationship between God and man. This spiritual awaking is the path to eternal happiness, peace and bliss, leading us to our ultimate destination — Moksha, Liberation, Nirvana or Enlightenment.

CHAPTER 3

COMPASSION

3.1   What is Compassion?

Compassion is a social feeling that motivates people to go out of their way to relieve the physical, mental, or emotional pains of others and themselves. Compassion is a sensitivity to the emotional aspects of the suffering of others. When based on notions such as fairness, justice, and interdependence, the compassion may be considered partially rational in nature. Compassion involves "feeling for another" and is a precursor to empathy, the "feeling as another" capacity (as opposed to sympathy, the "feeling towards another"). In common parlance, active compassion is the desire to alleviate another's suffering. Compassion involves allowing ourselves to be moved by suffering to help alleviate and prevent it. Other virtues that harmonize with compassion include patience, wisdom, kindness, perseverance, warmth, and resolve. The difference between sympathy and compassion is that the former responds to others' suffering with sorrow and concern whereas the latter responds with warmth and care. From the perspective of evolutionary psychology, compassion can be viewed as a distinct emotional state, which can be differentiated from distress, sadness, and love. The compassion consists of three facets: noticing, feeling, and responding".

The term "compassion" derives from the Latin word "compati", meaning "to suffer with". Compassion, not only entails being in contact with suffering, but also involves a profound commitment to relieve this suffering.

Compassion is, a synonym of empathic distress, which is characterized by the feeling of distress in connection with another person's suffering. This perspective of compassion is based on the finding that people sometimes emulate and feel the emotions of people around them.

Compassion is a sense of concern that arises in us in the face of someone who is in need or someone who is in pain. It is accompanied by a kind of a wishing (i.e. desire) to see the relief or end of that situation, along with wanting (i.e. motivation) to do something about it. Compassion is however not pity, neither an attachment, nor the same as empathetic feeling, nor even just simply wishful thinking. Compassion is basically a variation of love. Compassion starts with the feeling of alleviating the pain of others.

Identifying with another person is an essential process for human beings, something that is even illustrated by infants who begin to mirror the facial expressions and body movements of their mother as early as the first few days of their lives. Compassion is recognized through identifying with other people (i.e. perspective-taking), the knowledge of human behavior, the perception of suffering, transfer of feelings, knowledge of goal and purpose-changes in sufferers which leads to the absence of the suffering from others.

Compassion has three major requirements: (i) The compassionate person must feel that the troubles that evoke their feelings are serious; (ii) the belief that the sufferers' troubles are not self-inflicted; and (iii) the ability to picture oneself with the same problems in a non-blaming, non-shaming manner.

Compassion may induce feelings of kindness and forgiveness, which could give people the ability to stop situations that have the potential to be distressing and occasionally lead to violence.

Compassion has become associated with and researched in the fields of positive psychology and social psychology. Compassion is a process of connecting by identifying with another person. This identification with others through compassion can lead to increased motivation to do something in an effort to relieve the suffering of others.

In one study conducted by neuroscientists subjects' brain activities were recorded while they helped someone in need. It was found that while the subjects were performing compassionate acts, the caudate nucleus and anterior cingulate regions of the brain were activated, the same areas of the brain associated with pleasure and reward. One brain region, the subgenual anterior cingulate cortex/basal forebrain, contributes to learning altruistic behavior, especially in those with trait empathy. The same study showed a connection between giving to charity and the promotion of social bonding and personal reputation.

3.2   Difference between Compassion and Empathy

Compassion and empathy are fundamentally different but closely related. Consider these definitions:

Compassion is an emotional response to empathy or sympathy and creates a desire to help.

3.2.1   What is Empathy?

Empathy is our feeling of awareness toward other people's emotions and an attempt to understand how they feel. Empathy is an understanding of our shared humanity. It's the ability to see yourself in another person's shoes. Empathy is the awareness of other people's emotions and the ability to understand their feelings and emotions.

Empathy is deeply rooted in our brains and bodies. It's so rudimentary that it's actually instinctual. This type of empathy is what psychologists typically refer to as cognitive empathy. There are many reasons to practice empathy. It's both good for our personal health and our work relationships.

Empathy can make us unconsciously more sympathetic towards individuals we relate to more. This makes us less likely to connect with people whose experiences don't mirror ours.

3.3   Compassion and Self-Compassion

Compassion can not only be directed toward our loved-ones, but also to strangers and ultimately, to all human kind. Self-compassion is understood as the act of transferring these attitudes toward others, toward oneself, which, for many individuals is very hard. Self-compassion implies not to being judgmental toward oneself (on a cognitive level), and also being able to feel and connect with our own suffering (on an emotional level). Self-compassion means taking care of oneself as we would do for a loved one.

Self-compassion has three main components: (i) kindness, (ii) common humanity, and (iii) mindfulness. Common humanity means that all human beings are united by the experience of suffering. All humankind is united in the desire of being free of suffering. The concept of "common humanity", is similar to the notion of interconnectedness, which highlights that individuality and separation are false beliefs. This sense of interconnection is what frequently motivates compassion and compassionate behaviors. In contrast, the absence of common humanity would drive someone to isolation and to the denial of others' suffering. As a consequence of this, there is a barrier between "us" and "them": when we identify ourselves as being part of one group (us), there is an implicit separation with others (them). Finally, the third component of self-compassion is mindfulness. In the context of self-compassion mindfulness is close to decentering, as mindfulness refers to the capacity of feeling painful emotions and thoughts without over-identifying with them.

3.4 Assessment of Compassion and Self-Compassion

Currently, there are three questionnaires specifically designed to assess compassion.

3.4.1 The Compassion Scale (CS) (Pommier, E., Neff, K. D., Tóth-Király, I.)

Instructions:

Please read each statement carefully before answering. Indicate how often you feel or behave in the stated manner on a scale from 1 'Almost Never' to 5 'Almost Always.' Please answer according to what really reflects your experience rather than what you think your experience should be.

1.   I pay careful attention when other people talk to me about their troubles.

2.   If I see someone going through a difficult time, I try to be caring toward that person.

3.   I am unconcerned with other people's problems.

4.   I realize everyone feels down sometimes, it is part of being human.

5.   I notice when people are upset, even if they don't say anything.

6.   I like to be there for others in times of difficulty.

7.   I think little about the concerns of others.

8.   I feel it's important to recognize that all people have weaknesses and no one's perfect.

9.   I listen patiently when people tell me their problems.

10.   My heart goes out to people who are unhappy.

11.   I try to avoid people who are experiencing a lot of pain.

12.   I feel that suffering is just a part of the common human experience.

13.   When people tell me about their problems, I try to keep a balanced perspective on the situation.

14.   When others feel sadness, I try to comfort them.

15.   I can't really connect with other people when they're suffering.

16.   Despite my differences with others, I know that everyone feels pain just like me.

Coding scheme:

Kindness items: 2, 6, 10, 14

Common Humanity items: 4, 8, 12, 16

Mindfulness items: 1, 5, 9, 13

Indifference items (reverse-coded): 3, 7, 11, 15

To compute a total compassion score, take a grand mean of all items.

3.4.2   Self-Compassion Scale

The SCS was the first tool in this field. It was originally supposed to be a three-factor scale, but during its development, the authors realized that it should have six factors: the three main factors of compassion and their respective opposite constructs: kindness and self-judgment, common humanity and isolation, and mindfulness and over-identification. The SCS has 26 items, and each item is answered through a five-point Likert scale (0 being "nearly never" and 4 being "nearly always"). It offers both, a separate score of each component and a total score.

The SCS has adequate psychometric properties: high internal consistency (Cronbach's alpha=0.94 and good test–retest reliability (r=0.93. It demonstrated convergent validity with social connectedness and emotional intelligence, and divergent validity with self-criticism, depression, social desirability, narcissism, anxiety and rumination. In addition, several studies demonstrate good predictive validity, with positive correlations between self-compassion and mental health signs, such as lower depression and anxiety, and high happiness, optimism and life satisfaction.

3.4.3   Self-Compassion Scale Short Form

There is a shorter version of the SCS, the Self-Compassion Scale-Short Form (SCS-SF), with only 12 items. The SCS-SF demonstrated adequate internal consistency (Cronbach's alpha≥0.86) and a robust correlation with the SCS-Long form (r≥0.97).

3.5   Compassion in Buddhism

The first of the Four Noble Truths is the truth of suffering or dukkha (un-satisfactoriness or stress). Dukkha is one of the three distinguishing characteristics of all conditioned existence. It arises as a consequence of not understanding the nature of impermanence anicca (the second characteristic) as well as a lack of understanding that all phenomena are empty of self-anatta (the third characteristic).

When one has an understanding of suffering and its origins and understands that liberation from suffering is possible, renunciation arises. Renunciation then lays the foundation for the development of compassion for others who also suffer. This is developed in stages:

3.5.1   Ordinary compassion

The compassion we have for those close to us such as friends and family and a wish to free them from the 'suffering of suffering'.

3.5.2   Immeasurable compassion

This is the compassion that wishes to benefit all beings without exception. It is associated with both the Hinayana and Mahayana paths.

Itis developed in four stages called The Four Immeasurables:

- Loving kindness (Mettā),
- Compassion (Karuṇā),
- Joy (Mudita), and
- Equanimity (Upekshā)

3.6   Compassion in Hinduism

Yoga aims at physical, mental, and spiritual purification, with a compassionate mind and spirit being one of its most important goals. Various asanas and mudras are combined with meditation and self-reflection exercises to cultivate compassion.

In classical literature of Hinduism, compassion is a virtue with many shades, each shade explained by different terms. Three most common terms are daya, karuṇā, and anukampā. Other words related to compassion in Hinduism include karunya, kripa, and anukrosha. Some of these words are used interchangeably among the schools of Hinduism to explain the concept of compassion, its sources, its consequences, and its nature. The virtue of compassion to all living beings, is a central concept in Hindu philosophy.

In Mahabharata, Indra praises Yudhishthira for his anukrosha – compassion, sympathy – for all creatures. Tulsidas contrasts daya (compassion) with abhiman (arrogance, contempt of others), claiming compassion is a source of dharmic life, while arrogance a source of sin. Daya (compassion) is not kripa (pity) in Hinduism, or feeling sorry for the sufferer, because that is marred with condescension; compassion is recognizing one's own and another's suffering in order to actively alleviate that suffering. Compassion is the basis for ahimsa, a core virtue in Hindu philosophy and an article of everyday faith and practice. Ahimsa, or non-injury, is compassion-in-action that helps actively prevent suffering in all living things as well as helping beings overcome suffering and move closer to liberation.

Compassion in Hinduism is discussed as an absolute and a relative concept. There are two forms of compassion: one for those who suffer even though they have done nothing wrong and one for those who suffer because they did something wrong. Absolute compassion applies to both, while relative compassion addresses the difference between the former and the latter. An example of the latter include those who plead guilty or are convicted of a crime such as murder; in these cases, the virtue of compassion must be balanced with the virtue of justice.

The classical literature of Hinduism exists in many Indian languages. For example, Tirukkuṟaḷ, written between 200 BCE and 400 CE, and sometimes called the Tamil Veda, is a cherished classic on Hinduism written in a South Indian language. It dedicates Chapter 25 of Book 1 to compassion, further dedicating separate chapters each for the resulting values of compassion, chiefly, vegetarianism or veganism (Chapter 26), doing no harm (Chapter 32), non-killing (Chapter 33), possession of kindness (Chapter 8), dreading evil deeds (Chapter 21), benignity (Chapter 58), the right scepter (Chapter 55), and absence of terrorism (Chapter 57), to name a few.

Bhagavad Gita refers to 'Compassion' as below:

advesḥṭā sarva-bhūtānāṁ maitraḥ karuṇa eva cha
nirmamo nirahankāraḥ sama-duḥkha-sukhaḥ kshamī (12.13)

santuṣhṭaḥ satataṁ yogī yatātmā dṛiḍha-niśhchayaḥ
mayy arpita-mano-buddhir yo mad-bhaktaḥ sa me priyaḥ (12.14)

Meaning: Those devotees are very dear to Me who are free from malice toward all living beings, who are friendly, and **compassionate**. They are free from attachment to possessions and egotism, equipoised in happiness and distress, and ever-forgiving. They are ever-content, steadily united with Me in devotion, self-controlled, of firm resolve, and dedicated to Me in mind and intellect.

Compassion in Skanda Upanishad (1-5)

acyuto 'smi mahādeva tava kāruṇyaleśataḥ.
vijñānaghana evāsmi śivo 'smi kimataḥ param. . 1..

na nijaṁ nijavadbhāti antaḥkaraṇajṛmbhaṇāt.
antaḥkaraṇanāśena saṁvinmātrasthito hariḥ. . 2..

saṁvinmātrasthitaścāhamajo 'smi kimataḥ param.
vyatiriktaṁ jaḍaṁ sarvaṁ svapnavacca vinaśyati. . 3..

cijjaḍānāṁ tu yo draṣṭā so 'cyuto jñānavigrahaḥ.
sa eva hi mahādevaḥ sa eva hi mahāhariḥ. . 4..

sa eva hi jyotiṣāṁ jyotiḥ sa eva parameśvaraḥ.
sa eva hi paraṁ brahma tadbrahmāhaṁ na saṁśayaḥ. . 5..

Meaning (1-5): (Skanda says): Great god! Owing to an iota of your **compassion,** I am the lapse-less being (not lapsing from the identity). I am a mass of knowledge! I am also the Good – what more (can I need)? Owing to the waxing of the Internal organ, what is not spiritual appears as such; by its warning, this is nothing but pure knowledge or Hari. I am knowledge alone, unborn – what more? All that is other (than) It is inert and perishes like a dream.

He who discerns the consciousness as distinct from the inert is the unswerving mass of knowledge. Only he is Shiva, Hari, luminary of luminaries, the supreme god, the Brahman – I am that Brahman surely.

3.7   Compassion in Jainism

Compassion for all life, human and non-human, is central to the Jain tradition. To kill any person, no matter their crime, is considered unimaginably abhorrent. It is the only substantial religious tradition that requires both monks and laity to be vegetarian. It is suggested that certain strains of the Hindu tradition became vegetarian due to strong Jain influences. The Jain tradition's stance on nonviolence, however, goes far beyond vegetarianism. Jains refuse food obtained with unnecessary cruelty. Many practice veganism. Jains run animal shelters all over India.

At the core of the faith are the virtues of non-violence, compassion and respect for all living beings – from humans right down to plants and microorganisms.

The foundational premise in Jainism is ahimsa, which is translated as "non-violence". Jains avoid causing harm, whether intentional or unintentional, to any living being through their speech, actions or thoughts, choosing instead to practice compassion towards all life forms.

According to Jain scriptures, not just humans, animals and plants but also air, water, fire and earth contain living souls. Each of these souls is considered of equal value and should be treated with respect and compassion. Compassion plays a major role for the progress of the soul toward liberation. In other words, one is expected to behave in a way that does not cause discomfort and to do what one can to alleviate the pain and sufferings of others.

Jain scriptures require practicing Jains to not kill any living beings with two senses to five senses and minimize harm to earth, plants, water, air and fire (all the ekindriya jivas). Jains have gone to the extent of abandoning not just animal products but also any form of plant-based food which is considered ananthkay (root vegetables that have one body but contain many lives). Almost all our practices whether avoiding root vegetables or abstaining to eat after sunset are based on compassion towards other living beings. Eating at night is prohibited because germs that we cannot see spread rapidly at night so after sunset, proper food does not enter the stomach.

Compassionate living is not just restricted to food, it encompasses every element that makes up our world. How does one practice compassionate living? While Jainism teaches us the principles, the application is something that should change as our lifestyle changes. With the burgeoning population and industrialization, the need to make the practice of compassion and non-violence broader in our lives is more pertinent than ever. Without our knowledge, many of our actions in the current era are inflicting immense impact on the living beings as well as ekindriya jivas. It is possible to avoid or minimize such harm when we are informed about the impact of our action on other living beings.

Compassionate practices in Jainism go beyond food. Many products used in our daily lives are derived from animals. These include leather, silk, and wool. The technology has developed enough these days to produce functional alternatives. Yes, some of these might be inferior in some respects – e.g., PU Leather often suggested as an alternative to natural leather may degrade faster, but we do need to keep it in the context of the harm we would inflict on animals if we continued using leather. Similarly for silk, where more than 6000 worms are killed in boiling water just to produce one silk saree – again an area where there are a lot of cruelty free alternatives available.

CHAPTER 4

EGO DEATH AND EGO DISSOLUTION

4.1 What is Ego Death?

To understand ego death, first you have to understand the ego. The ego is a sense of self that you develop at a very young age. It relates to your feelings about your own importance and abilities. It is how you identify and what you identify with, whether it is a gender identity, your beliefs or your values. You can think of your ego as the image you hold of yourself, which affects everything from how you perceive the world to how you behave around others. Our ego (or personal identity) can keep us disconnected from reality and, instead, perceive life through the lens of duality. This causes us to see everything as either right or wrong, good or bad.

Ego death is a "complete loss of subjective self-identity". The term is used in various intertwined contexts, with related meanings. Jungian psychology uses the synonymous term psychic death, referring to a fundamental transformation of the psyche. In death and rebirth mythology, ego death is a phase of self-surrender and transition.

Ego death is the (often instantaneous) realization that you are not truly the things you've identified with, and the "ego" or sense of self you've created in your mind is a fabrication. In some instances, it can offer a profound feeling of peace and connectedness with all that is, as the walls of separation the ego creates come crumbling down.

Ego death is better described as an awakening or realization that will transform your life for the better. When we experience the loss of our ego, we can connect to our true nature, others, and the world around us. We are no longer held hostage to the ego's power that keeps us feeling small, alone, and in many cases, angry.

Ego death triggers self-discovery, where we discover our true nature, which means who we are on a soul level. Through this mystical experience, we can understand our soul's purpose and gain new perspectives on life. In this sense, our ego must die for this part of us to be born.

The concept is also used in contemporary New Age spirituality and in the modern understanding of Eastern religions to describe a permanent loss of "attachment to a separate sense of self" and self-centeredness. This conception is an influential part of Eckhart Tolle's teachings, where Ego is presented as an accumulation of thoughts and emotions, continuously identified with, which creates the idea and feeling of being a separate entity from one's self, and only by disidentifying one's consciousness from it can one truly be free from suffering.

Ego death and the related term "ego loss" have been defined in the context of mysticism by the religious studies as "an imageless experience in which there is no sense of personal identity. It is the experience that remains possible in a state of extremely deep trance when the ego-functions of reality-testing, sense-perception, memory, reason, fantasy and self-representation are repressed. Muslim Sufis call it fanaa ('annihilation'), and medieval Jewish kabbalists termed it 'the kiss of death'".

In Jungian psychology, ego death is defined as "a fundamental transformation of the psyche". Such a shift in personality has been labeled an "ego death" in Buddhism, or a psychic death by Jung.

In psychedelic culture, ego death or ego loss is defined as part of the (symbolic) experience of death in which the old ego must die before one can be spiritually reborn. The ego loss is defined as "… complete transcendence – beyond words, beyond spacetime, beyond self. There are no visions, no sense of self, no thoughts. There are only pure awareness and ecstatic freedom".

Several psychologists working on psychedelics have defined ego-death as " the loss of ego-feeling". It is a sense of total annihilation. This experience of "ego death" seems to entail an instant merciless destruction of all previous reference points in the life of the individual. Ego death means an irreversible end to one's philosophical identification with "skin-encapsulated ego". The temporary ego death is loss of the separate self, or, in the affirmative, a deep and profound merging with the transcendent other.

4.2   Ego Death: Views of Spiritual Traditions

4.2.1   Ego Death in Buddhism

In Zen practice, Ego-death is also called "great death", in contrast to the physical "small death". The ego death that Buddhism encourages makes an end to the "usually-unconsciousness-and automated quest" to understand the sense-of-self as a thing, instead of as a process. Accordingly, the meditation is learning how to die by learning to "forget" the sense of self.

The ego-lessness appears in the gaps and spaces between thoughts, which usually go unnoticed. Existential anxiety arises when one realizes that the feeling of "I" is nothing more than a perception. Only egoless awareness allows us to face and accept death in all forms.

4.2.2 About Ego in Bhagavad Gita (3.27)

*prakriteh kriyamāṇāni guṇaih karmāṇi sarvaśhah
ahankāra-vimūḍhātmā kartāham iti manyate*

Meaning: All activities are carried out by the three modes of material nature. But in ignorance, the soul, deluded by false identification with the body, thinks of itself as the doer.

4.2.3 Ego in Maha Upanishad

*ahaṃkāramayīṃ tyaktvā vāsanāṃ līlayaiva yah.
tiṣṭhati dhyeyasaṃtyāgī sa jīvanmukta ucyate. . II-45..*

Meaning: II-45. He is said to be 'Liberated while living', Who is silent, **egoless,** prideless, avoiding jealousy and does actions without agitation.

*ahaṃkāravaśādāpadahaṃkārāddurādhayah.
ahaṃkāravaśādīhā nāhaṃkārātparo ripuh. . III-16..*

*ahaṃkāravaśādyadyanmayā bhuktaṃ carācaram.
tattatsarvamavastveva vastvahaṃkārariktatā. . III-17.*

Meaning: III-16-17. From **ego** does danger arise, so do bad mental ailments and desire – there is no enemy more dangerous than **Ego**; whatever in the moving and unmoving world was enjoyed by **Ego** – all that is unreal; only freedom from **Ego** is real.

4.2.4 Ego Death Sufism (Fanaa)

Fanaa is the Sufi term for extinction. It means to annihilate the self, while remaining physically alive. Persons having entered this state are said to have no existence outside of, and be in complete unity with, Allah. Fanaa is equivalent to the concept of nirvana in

Buddhism and Hinduism or moksha in Hinduism which also aim for annihilation of the self.

Abu Yazid al-Bistami approached the Divine Presence and "knocked on the gate". He was asked, "Who is there?" "I have come, Oh my Lord", replied Abu Yazid. He was told: "There isn't any place here for two. Leave your ego behind and come". When Abu Yazid once again approached the Divine Presence and was asked who it was, he said: "You, oh Lord".

The "annihilation of the self" (fanaa fi 'Allah') refers to disregarding everything in this world because of one's love towards God. When a person enters the state of fanaa it is believed that one is closest to God.

The Qalaba (heart) is sandwiched between the nafs (EGO) and the Rooh (SOUL) The entire objective of annihilation is to destroy the nafs to that Heart can recognize the soul. Sufis say that the soul has the spark of divine as in Quran, all souls come from God.

The nature of fanaa consists of the elimination of evil deeds and lowly attributes of the flesh. In other words, fanaa is abstention from sin and the expulsion from the heart of all love other than the Divine Love; expulsion of greed, lust, desire, vanity, show, etc. In the state of fanaa the reality of the true and only relationship asserts itself in the mind. One realizes and feeds that the only real relationship is with Allah Ta'ala, fanaa means to destroy yourself. if you destroy yourself in the love of Allah then that fanaa will convert into entire life means *abdi zindgi*, and for that one you have to destroy your will and yourself on the will of Allah.

In the death of the ego love is born, God is born, light is born. In the death of the ego you are transformed; all misery disappears as if it had never existed. Your life right now is a nightmare. When the ego dies nightmares disappear and a great sweetness arises in your being, and a subtle joy, for no reason at all. Beyond this is the stage of intimacy (uns) at which the immanence of the Lord is perceived:

"And I am closer to man than his jugular vein" QURAN;

On the path of ego annihilation; Bayazid said:

I became like an iron master for twelve years. I put my nafs and ego in the stove of discipline, and prepare it with the fire of striving, mold it on the platform of remorse, hammer it with regret until my nafs became my mirror. I was my own mirror for five years. Until one day when I thought I was the greatest among great learned. As soon as this thought came to my mind, I packed up and went to Khorasan. I stayed in a shelter and promised myself that I would not leave this place unless I receive a message from Allah. On the fourth day I saw a camel rider coming towards me. A thought passed my mind that I could stop that camel right there. The rider looked at me and said: Do not make me to destroy Bastam and Bayazid altogether. I lost my senses. When my senses came back to me I asked him: Where are you coming from? he said: From the side where your promise is kept. The he said: Bayazid, keep and protect your heart; then he left. It is said after this incident whatever passed through Bayazid 's mind would appear in front of him.

What Bayazid is explaining in this story is that;

He who recognizes himself.. recognizes GOD;

In Bayazidian Sufism, one has to get rid of the pseudo-personality that one has created for oneself. We all want to be accepted and respected by others. Most of the time we are led by society and our own cultural norms to create a false sense of ourselves.

Whenever you are now and here, there is no ego to be found. You are a pure silence. Ego is the centre of the false mind.

Your ego is your hell, your ego is your misery, your ego is the cancer of your soul. When desires cease, the other world opens. The other world is hidden in this world. But because your eyes are full of desires, full of the ego, you cannot see it.

This was the lost secret of the ages.. the divine lies within you, if only you would listen to it; all mystical paths really aim to remove ego from the self so only name of God remains.

God tells Bayazid that He doesn't care if he sees the world or not. He only cares if Bayazid doesn't see himself. And it's only when he ceases to see himself that Bayazid can truly say that he has seen God.

This is when Bayazid says:

The final clue is buried in your stare;

So long as 'I' continues to exist

The sun I seek is shrouded in "I's" mist

Bayazid repents first from thinking he has seen God, and second he repents from that repentance for this is just another manifestation of his being; finally, he repents from seeing his own existence altogether.

He addresses GOD; "Oh, Allah, this is how I see myself. I am not offering You my life's mortification, my constant prayers, my day and night fasting, You know that nothing will take me from You. I confess that I am shameful, I have nothing, You are the One who has given me all this fortune. I witness that there is no god but You. You have accepted me. Purify me from my errors, forgive my faults, wash away my shortcomings.

4.3   Unity and Ego Loss

That experience can be profound and positive, fostering a sense of unity with other human beings, nature and the universe. Or it can sink us into painful isolation, estranging us from our surroundings, circumstances and identities — the proverbial "bad trip."

Ego loss consists of negative experiences such as depersonalization, detachment from reality and dissociation, and even psychosis, while unity carries a mystical sense of oneness with a larger whole, he explained. Unity experiences are associated with mindfulness, positive emotions and an extraverted personality, which naturally seeks connections with others.

The researchers also explored variables associated with ego dissolution. A tendency to be mindful in everyday life, for example, was negatively related to ego loss, but positively associated with unity. A predisposition to hallucinations is associated with greater susceptibility to ego dissolution as a whole, as well as the negative experience of ego loss — which raises interesting questions as to peoples' susceptibility to "bad trips."

4.4   Ego Dissolution Scale

Ego dissolution (also called ego loss, ego disintegration, ego death, or self-loss) refers to changes in information processing regarding the self and to concomitant alterations in the experience of the self that are linked specifically with the sense of diminution, loss, or disintegration of the self.

Ego dissolution is a multifactorial and multidimensional construct that encompasses both positive and negative experiences and can be conceptualized, manifested, and assessed in different ways and studied in response to diverse antecedents (e.g., psychedelic drug use, meditation). One researcher contended that

"...no characterization subsumes and epitomizes the psychedelic experience more precisely than the fundamental alteration of self, known as ego dissolution". Another researcher noted the positive transformational qualities of ego dissolution in observing that it can be a valuable self-enhancing experience in response to psychedelics and other hallucinogens (e.g., dissociative anesthetics, agonists of the kappa opioid receptor), but the experience can also "be terrifying or blissful depending on the context and previous experience of individuals. In Buddhist contemplative philosophy, the achievement of anatman, or "no-self," represents a crucial insight. Dismantling the "illusion of the self" is conceptualized as a critical step in transcending suffering and thus is often targeted in meditative practices as a highly desirable milestone. Studies using Dittrich's Abnormal Mental States (APZ) questionnaire and its predecessors have further examined diverse states of ego dissolution associated with intense meditation, psychedelic use, and other anomalous experiences, including mystical-type experiences and "anxious ego-dissolution," which refers to a decidedly negative experience related to derealization.

A researcher (2017) captured the multifactorial nature of ego dissolution in identifying two components of STEs: an annihilational component and a relational component. The "annihilational component" is evident when the subjective sense of the bodily self, self-boundaries, social boundaries, and self-salience dissolve. This component can be associated with negative experiences, such as psychotic and delusional episodes, schizophrenia, and depersonalization, but also with positive aspects of ego dissolution in the case of diminishing excessive self-focus in anxiety and depression.

In alternate states of ego dissolution, the boundary between subjective experiences and the sense of a separate self, who witnesses the constant ebb and flow of experiences, can become translucent to the point of evoking a positive and potent sense of self-world unity, oneness with the universe, and connection with living beings. The researchers describe this second relational component of ego dissolution as a sense of connectedness or unity beyond the self, which has been linked with mystical-type experiences related to positive enduring therapeutic effects of ego dissolution.

Some researchers found that only five items loaded onto a common factor that they interpreted as ego dissolution and that also correlated with mystical experiences. Clearly, research is needed to further clarify and refine the measurement and the nature of the construct of ego dissolution in the context of a valid and internally consistent scale designed to do so.

A reliable scale that was developed based on diverse items pertaining to alterations in the perception of the self. The intention in doing so was twofold: (1) to assess trait-like aspects of ego dissolution, which have not received attention in previous research and (2) to facilitate future research on ego dissolution in casting a broad net of measures that we argue are potentially related to ego dissolution. In doing so, specific hypotheses were addressed and research questions regarding the correlates of ego dissolution and the convergent and discriminant validity of this measure.

Another important aim of our study was to examine the extent of linkage of measures of ego dissolution and dissociation in an exploratory manner. We predict that ego dissolution and dissociation measures will be highly correlated yet potentially distinct and that domain-specific variance in ego dissolution will associate uniquely with a more expansive nomological network of variables we selected on a theoretical and empirical basis. Beyond indices of dissociative experiences, we included an array of variables that we predict will go some way toward demarcating the domain of ego dissolution, as well as measures that have not yet been assessed in mapping the breadth of the construct.

Although a scale exists to measure ego dissolution in response to an antecedent event, no current measure exists to assess a propensity towards ego dissolution more generally. Thus, in a research it aims to (a) develop a scale that facilitates the study of trait-like aspects of ego dissolution and (b) evaluate the validity of scale and its internal consistency, factor structure, and indications of convergent and discriminant validity.

In a survey, 572 undergraduate students were recruited (61.6% female, 37.5% male, 0.2% intersex; Mage = 18.90, SD = 1.17) from an online participant pool in exchange for course credit. The majority of the sample identified as Caucasian (66.1%), 19.5% identified as Asian, 7.9% identified as Black/African American, 5.4% endorsed multiple identities or reported their identity was not listed, 0.7% identified as Native American/Native Alaskan, and 0.4% identified as Native Hawaiian/Pacific Islander. A small subset

The goal was to construct a valid and internally consistent scale of ego dissolution with strong psychometric properties. To do so, 14 items were judged to represent the content domain of ego dissolution to an EFA. Bartlett's test of sphericity was significant ($\chi2$ (45) = 1674.328, p <. 001), suggesting that factor analysis was an appropriate statistical procedure. The KMO measure of sampling adequacy was 0.85, suggesting good sampling adequacy.

CHAPTER 5

SPIRITUAL MATURITY AND STABILITY

5.1   What is Spiritual Maturity?

Spiritual Maturity is:

(i)   when you stop trying to change others, …instead focus on changing yourself.

(ii)   When you accept people as they are.

(iii)   When you understand everyone is right in their own perspective.

(iv)   when you learn to "let go".

(v)   when you drop "expectations" from a relationship and give for the sake of giving.

(vi)   when you understand whatever you do, you do for your own peace.

(vii)   when you stop proving to the world, how intelligent you are.

(viii)   when you don't seek approval from others.

(ix)   when you stop comparing with others.

(x)   when you are at peace with yourself.

(xi)   when you are able to differentiate between "need" and "want" and are able to let go of your wants.

(xii)   when you stop attaching "happiness" to material things.

5.2   Stability of Mind according to the Bhagavad Gita

With our minds, we can fathom the mysteries of the world and the universe and make informed decisions. Although our minds are the highest expression of Nature in the mortal world, they are inherently unstable and subject to several modifications and afflictions. Hence, we cannot always rely upon them to find right solutions or think correctly. Many factors contribute to their instability. Unless they are neutralized we do not experience peace. The world itself is both the cause and the effect of the human mind. There is an inseparable connection between the two. It is difficult to say where one ends, and the other begins. We recreate the world in our minds according to our own desires, knowledge and expectations while we superimpose our minds upon the world to make sense of it. In other words, your mind is shaped by your perception of the world, while your perception of the world is in turn shaped by your mind. In both situations the mind experiences modifications or disturbances.

The instability of the human mind or its restlessness is not an aberration or an abnormality but a natural condition of the human personality. Indeed, it is an existential imperative to manage the complexity of our world and the problem of survival. It helps the mind efficiently process vast amounts of perceptual data and remain wakeful to the threats that are present in the environment. According to Hinduism, the instability of the mind is designed by Nature or Prakriti as part of its Maya to keep the beings deluded and distracted so that they remain outwardly focused with the distractions and attractions of the world and become oblivious of their spiritual nature.

Nature accomplishes its designs with the help of the triple Gunas (modes or tendencies) namely Sattva, Rajas and Tamas. Sattva drives beings to seek peace and happiness, Rajas drives them to engage in egoistic actions to seize and control things, while Tamas tends to induce inertia in them to make them slothful and ignorant. The three are responsible for desire-ridden actions in all beings, which lead to their suffering, rebirth and bondage in the mortal world.

Understanding the modifications of the mind and silencing them is the central feature of the Yoga tradition and Hindu spirituality. The ascetic traditions of Hinduism aim to stabilize the mind to experience self-realization. One may also notice that the entire discourse of Lord Krishna was in direct response to the problem of Arjuna's disturbed state of mind and to prepare him spiritually for the impending war. In the following discussion, we focus upon the problem of the fickle nature of the mind and its resolution according to the Bhagavad Gita.

The Bhagavad Gita recognizes suffering as the central problem of human life and traces it directly to the influences to which the mind is subject. It suggests that the movements of our senses among sense-objects are responsible for the disturbances of our minds. Their repeated contact with the sense-objects leads to attraction and aversion, which create desires and attachments, and which in turn influence our thoughts, emotions and actions. When we perform actions under their influence, we become subject to karma, which in turn leads to bondage and suffering. In other words, desires and craving are at the root of the problem. They keep our mind disturbed and distracted. Therefore, if we want to resolve the problem of mental instability and find peace and equanimity, we must first focus upon our desire-ridden actions and cultivate sameness. The Bhagavad Gita identifies the following chief causes of the fickleness of the human mind.

5.2.1 Dualities of Life

As stated before, attraction and aversion to the dualities or opposites of life lead to the conflicting states of happiness or unhappiness, pleasure or pain, and suffering or enjoyment. When we are caught in them, we experience emotional highs and lows and do not experience peace. Of them, the duality of the subject and the object is the most difficult to overcome.

5.2.2 Desires and Attachment

Because of our limited abilities and egoism, we are subject to numerous desires. Whether they are fulfilled or not, desires have a tendency to keep the mind disturbed as you become caught between positive and negative emotions. Therefore, peace is rarely achieved by seeking things or fulfilling desires.

5.2.3 Impurities

The mortal world is an impure world since it is filled with the triple gunas and the influence of Maya. So is the case with our minds and bodies. Because of the impurities such as egoism, desires, ignorance and delusion we cannot truly discern things or make right decisions, whereby we make mistakes and subject ourselves to physical and mental suffering.

5.2.4 Karma

The consequences of desire-ridden actions result in karma, which is cumulative and continuous. The unexhausted karma of our past lives becomes carried forward to the current life and influences our thinking and actions. It is chiefly responsible for our inexplicable behavior, predominant desires and habitual thoughts, which keep the mind fickle and disturbed.

5.3 The Bhagavadgita on the Problem of Fickle Mind

The Bhagavad Gita recognizes the mind as one of the finite realities (tattvas) of Nature and part of the Field (Kshetra). It is the lord of the senses and the sixth sense. However, since it is directly connected to the senses, it remains deluded and disturbed and cannot focus upon anything for long. When the mind is disturbed, one cannot withdraw the senses from the external world or become disengaged from its numerous distractions.

According to the Bhagavad Gita, the mind is fickle by nature, like wind, candle light or water in motion. It is unstable because of the activity of the senses and the desires and attachments which they induce. An unstable mind is verily the cause of delusion, an enemy of the self (6.6), whereas a stable mind is the very seat of Supreme consciousness, the doorway to self-realization and a precondition for the attainment of immortality.

bandhur ātmātmanas tasya yenātmaivātmanā jitaḥ
anātmanas tu śhatrutve vartetātmaiva śhatru-vat. (BG, 6.6)

Meaning: For those who have conquered the mind, it is their friend. For those who have failed to do so, the mind works like an enemy.

A fickle minded person loses his discernment and falls into evil ways. He remains deluded and bound to the cycle of births and deaths.

5.3.1   The importance of Stable Mind and Intelligence

One of the benefits of mental stability is clarity in thinking. When the mind is calm, it thinks clearly and lets the intelligence shine. Intelligence (buddhi) is important to thinking and discernment. It is the highest faculty of the mind. A yogi whose mind is stable and whose intelligence is sharp and focused is described in the Bhagavad Gita as sthithaprajna (a person with established intelligence or reason). With it, one can clearly discern things, know the truth from falsehood, reality from delusion and escape from the confusion and delusion of Maya.

The Bhagavad Gita declares that a person is said to be stable of mind when he renounces all desires and remains satisfied in the Self by the Self, remaining the same to all the dualities such as heat and cold, pleasure and pain, sorrow and happiness, friend and foe, success and failure, respect and disrespect, himself and others, and so on. He remains undisturbed and unmoved by the vagaries of the world or his own changing conditions.

From our own experience we know that we suffer mainly because of our likes and dislikes and our desire to have or not to have what we like or dislike. When things do not happen according to our expectations or when we find ourselves in adverse situations, we experience negative emotions and suffering. A person of discerning intelligence knows that he cannot control the world (adhibhautika) or the acts of God (adhidaivika) but he can control himself and his inner world (adhyatmika) through self-discipline and experience peace.

5.3.3 Achieving Mental Stability

The truth is we have little control over the world and things that are external to us. We cannot control the world, but we can control our inner world, thoughts, reactions, attitude, feelings and emotions. We can learn to control our desires and our emotional reactions and responses to the happenings and circumstances in our lives. We can change our perspective and how we interpret our experiences and perceptions. Most importantly, with effort we can cultivate detachment and not respond to any situation at all.

The Bhagavad Gita points to the same approach. It suggests that the mind can be controlled through "abhyas" (practice) and vairagya (dispassion), by withdrawing the senses from the sense-objects the way a tortoise withdraws its limbs, overcoming desires through detachment from the sense objects, living in solitude, free from possessiveness, and by fixing our minds constantly upon

God. In other words, a change in the orientation and direction of the mind, from the outward obsession with worldly things to the inward focus upon the hidden truth, and from objectivity to subjectivity.

However, they are not the only means to reorient the mind. Through physical and mental purity, devotion, concentration, desireless actions, renunciation of the fruit of actions, living in solitude, accepting life as it unfolds, complete surrender and living life as a sacrifice one can rein the mind and stabilize it in the contemplation of the Self. Moderation in thinking and action is another important practice suggested by it (6.16).

nātyaśhnatastu yogo 'sti na chaikāntam anaśhnataḥ
na chāti-svapna-śhīlasya jāgrato naiva chārjuna. (BG, 6.16)

Meaning: O Arjun, those who eat too much or too little, sleep too much or too little, cannot attain success in Yoga.

The mind becomes stable when one realizes how the gunas delude beings and transcends them (14.24-25).

sama-duḥkha-sukhaḥ sva-sthaḥ sama-loṣhṭāśhma-kāñchanaḥ
tulya-priyāpriyo dhīras tulya-nindātma-sanstutiḥ (BG., 14.24)
mānāpamānayos tulyas tulyo mitrāri-pakṣhayoḥ
sarvārambha-parityāgī guṇātītaḥ sa uchyate (BG., 14.25)

Meaning (BG: 14.24-25): Those who are alike in happiness and distress; who are established in the self; who look upon a clod, a stone, and a piece of gold as of equal value; who remain the same amidst pleasant and unpleasant events; who are intelligent; who accept both blame and praise with equanimity; who remain the same in honor and dishonor; who treat both friend and foe alike;

and who have abandoned all enterprises – they are said to have risen above the three guṇas.

5.4 Methods to Restrain the Mind

The following are the important methods which are suggested by the scripture to deal with the impurities of the mind and stabilize it in the Self (Atma) through controlled concentration (samyama) which lead to oneness.

(i)   Restraining the mind and the senses through self-discipline;

(ii)   Overcoming desires by cultivating contentment;

(iii)   Practicing detachment and dispassion;

(iv)   Renouncing the desire for the fruit of one's actions;

(v)   Cultivating sattva or purity;

(vi)   Practicing discernment so that one can overcome delusion;

(vii)   The pursuit of the knowledge of liberation through learning and study;

(viii) Performing selfless actions as an offering to God;

(ix)   Devotional and contemplative practices which lead to equanimity and self-absorption; and

(x)   Cultivating sameness to the pairs of opposites.

These practices cleanse the mind and intelligence and stabilize them in the Self. A stable mind is the foundation of liberation. Through renunciation and detachment and engaging in selfless actions as an offering to God, one can cross the ocean of mortal life on the raft of devotion with the mind constantly engaged in the contemplation of God or Self.

The practice is arduous, but even a little effort in the right direction has its beneficent effect since there is no loss on the path. If a devotee stumbles and fails to achieve liberation, he may still hope to continue on the path in the next life and begin from where he left. However, it is important not to fall into evil ways or commit grave sins through self-destructive actions. One should remain a friend of the Self and not become its enemy. Lord Krishna says that sinful souls who succumb to demonic qualities will be cast into the darkest hells and sinful wombs, from where their chances of recovery are very slow.

5.4.1 The state of Stable Mind (sthitha-prajna)

The Bhagavad Gita describes the positive outcomes of the various yogas which lead to mental stability. With mental stability comes undisturbed peace and unending calm. All sorrows cease to bother the person of stable intelligence as he does not crave for anything, nor does he complain about anything. With discretion he makes right choices and guides his life to liberation, without falling into the trap of the senses or the dualities of life.

Cultivating sameness towards the pairs of opposites and overcoming attraction and aversion, he remains satisfied with whatever that happens to him by the will of God (yaddruchcha labha samtushta) and goes through all experiences with the same attitude. With discernment and detachment, he becomes immune to the play of Prakriti, knowing that desire-ridden actions are caused by the gunas and he is not the one who is causing them.

Having restrained his mind and body and having conquered desires and attachments with discerning intelligence, he goes beyond the sense of duality (dwandatitha) and overcomes jealousy, anger, egoism and such other evils. By cultivating higher knowledge through self-purification, selfless actions, devotional services and contemplative practices, he lifts the veil of illusion covering his mind and sees the beauty and splendor of his true Self which is hidden deep within himself. He goes beyond objectivity and mental modifications to become absorbed in the Infinite Consciousness and sees himself in all and all in himself.

For that person there is no rebirth and no suffering. He sees the world as his own extension. Upon his death, he travels by the path of the Sun to the highest world of Brahman, from where he never returns.

CHAPTER 6

SPIRITUAL ACCEPTANCE AND SURRENDER

6.1 Acceptance as a Step Toward Spiritual Intelligence

Just as there are laws of the physical universe such as Newton's laws of gravitation and motion, there are five sequential rules that govern spiritual consciousness. They are acceptance, cooperation, understanding, loving, and enthusiasm. Each one is a doorway to the next. Our awareness of the presence of spirit in our lives is governed by these five laws. As we become able to comprehend and align ourselves with them, we gain access to the treasures they guard.

The spiritual consciousness is our ability to know our divine nature and to let that inform how we function in our lives. No matter what one's beliefs are regarding God, spirituality, or religion, the laws of spirit represent a passageway to mental and emotional freedom. When we do not work in cooperation with these laws, our consciousness typically operates in a reactive mode to external conditions — perceiving ourselves to be victims or winners in the game of life.

The first law of spirit, or stepping stone along this path to freedom, is **acceptance**. Real acceptance is not for wimps, nor is it a wishy-washy passive way of making do with whatever is present. It is not a "whatever" attitude of resignation either. **Acceptance is a conscious choice to drop all forms of resistance** to whatever has come present in the moment and making the most of it. Acceptance isn't about liking or approving of something. It is about letting life flow and unfold without getting in the way. It is about being receptive rather than exerting resistance to what comes present. Instead of focusing on the past or the future or wanting things to be different than they are, we open to **what is true in the moment**. This absence of "against-ness" allows us to engage our reality in such a way that we can learn from it and strengthen our ability to function in this world.

For many of us, our first impulse is to resist something that we do not like that comes our way. Acceptance requires overriding this impulse and choosing to breathe into and through the experience, trusting that it has a value that is for us and not against us. The truth of the matter is that **resistance prolongs the negative experience, and acceptance allows for the possibility of changing our experiences by changing our attitudes**.

Consider the bride who had her heart so set on having an outdoor wedding that she didn't make a solid plan B in case of inclement weather. As her wedding day arrived, the storm clouds were rolling in and the forecast was not promising. She was the one who would make the final judgment call about whether or not to move the ceremony indoors. She woke up worrying about the weather, and was distracted by her concerns throughout the entire day leading up to her 6 p.m. ceremony. She missed out on all the available joy and excitement of spending her wedding day with her bridal party. She didn't realize that while the impending storm was out of her control, the internal weather in her heart and mind was entirely up to her. Instead of deciding to play it safe and give her team time to set up the ceremony indoors, she waited until the very last minute. She was sitting there sobbing in her wedding dress, with makeup dripping down her stubborn face. Just then, the sky blackened and there was a torrential downpour. All the chairs were soaked, the guests ran for cover in a panic and the staff was out in the rain gathering the chairs, whisking them into the reception tent, toweling them dry while sliding over the wet and treacherous floor. Imagine how different this bride's day would have been if she had been able to accept that she had no control over the weather.

Here are two things to watch out for next time you have preconceived notions about how you want things to be and then reality presents you with something very different:

If you hold on to your desire, you create a mental and emotional tension between your desire and reality. The distance between the two will be connected by a stream of negative emotions. Alternatively, if you choose to accept reality, all that energy that would have been spinning around in negativity can be invested in figuring out how to make the best of your situation.

As in the example of the bride described above, no amount of wishful thinking or attachment to a desired outcome makes it any more probable than it is. Consider only truly viable alternatives. If there is a 70 percent chance of rain, attaching yourself to the idea of sunshine merely sets you up for unhappiness and blinds you from seeing, appreciating, and investing in the real options that are available to you.

Ultimately, acceptance is about trusting yourself to rise to whatever occasion presents itself to you. It is about being open to ALL of life, knowing that it all has value whether you like it or not.

The reason more of us are not spiritually aware people is that we often don't or won't accept what is happening. Acceptance is a flow of consciousness that continually moves on to the next thing.

So accept whatever comes your way, and don't grumble against anything that happens to you. If it happens, it happens. Go on about your business. Keep flowing. You cannot control circumstances from the outside, so instead of resisting pain or failure and defending against it, you can embrace and encompass your pain and your failures, fully accepting them so that they become part of you.

You then can let them go because they are part of your inner environment — they are within your domain — and the loving of your Soul can dissolve them.

Accept — then act. Whatever the present moment contains, accept it as if you had chosen it … This will miraculously transform your whole life.

6.1.1   Acceptance as a Spiritual Practice

As you commence on your spiritual journey, embracing acceptance as a practice can lead you to profound growth. By embracing the present moment, letting go of resistance, and surrendering to the divine, you open yourself to a deeper connection with the universe. Through acceptance, you can cultivate peace, understanding, and harmony within yourself and with the world around you.

6.1.2   Embracing Present Moment

Have you ever truly embraced the power of living fully in the present moment, allowing acceptance to guide your spiritual practice? Embracing the present moment is a profound way to cultivate inner peace and set out on a journey of inner transformation.

Here are three ways to help you embrace the present moment:

(i) Practice Mindfulness

Engage in mindfulness activities to anchor yourself in the now and appreciate the beauty of each moment.

(ii) Let Go of Worries

Release worries about the past or future, and focus on what you can control in the present.

(iii) Express Gratitude

Take time to acknowledge and express gratitude for the simple joys that surround you, fostering a deeper connection to the present.

6.1.3   Letting Go Resistance

To truly deepen your spiritual practice, consider the profound power of letting go resistance and embracing acceptance as a transformative way of being. Releasing control isn't about giving up but understanding that some things are beyond our influence. When you resist change, you create unnecessary suffering. By accepting impermanence, you open yourself to growth and new possibilities.

Life is in constant flux, and when you cling to what should be, you miss the beauty of what is. Letting go of resistance allows you to flow with the rhythm of life, finding peace in the midst of chaos. Embrace the wisdom that comes from surrendering control and acknowledging the transient nature of existence.

6.1.4   Surrendering to Divine

Embrace the transformative power of surrendering to the divine, allowing acceptance to guide your spiritual journey towards inner peace and enlightenment. When you practice trustful surrender, you open yourself up to profound spiritual growth and connection.

Here are three ways surrendering to the divine can positively impact your life:

(i) Letting go of control

Surrendering allows you to release the need to control every aspect of your life, leading to a sense of freedom and peace.

(ii) Deepening Faith

Trustful surrender strengthens your faith in a higher power, fostering a sense of trust and security in the universe's plan for you.

(iii) Finding Inner Peace

By surrendering, you invite a sense of calm and serenity into your life, helping you navigate challenges with grace and resilience.

(We shall discuss more about the 'Spiritual Surrender' subsequently in this chapter).

6.1.5 Finding Peace Through Acceptance

In the journey to finding peace through acceptance, you must first learn to let go of resistance and open your heart to the flow of life. Inner peace awaits those who embrace acceptance, allowing tranquility to blossom within. When you release the need to control every outcome and instead trust in the universe's unfolding, you invite serenity into your being.

Essential compassion plays a crucial role in this process. Be gentle with yourself as you navigate challenging circumstances. Acknowledge your emotions without judgment, offering kindness and understanding to your inner self. Through self-compassion, you create a nurturing environment for acceptance to take root.

Embrace the present moment with all its imperfections and uncertainties. Life is a tapestry of experiences, each thread contributing to the beauty of the whole. By accepting what is, you free yourself from the burdens of resistance, paving the way for peace to settle within your soul. Trust in the power of acceptance to bring you the peace you seek.

6.1.6 Acceptance and Resilience

By cultivating a mindset of acceptance, you strengthen your resilience in the face of life's challenges, empowering yourself to navigate adversity with grace and fortitude. Embracing acceptance opens the door to profound growth and transformation within you.

Here are three ways resilience and growth are intertwined with acceptance:

(i) Embracing Change

Acceptance allows you to adapt to unexpected changes by acknowledging them without resistance, fostering resilience in the face of uncertainty.

(ii) Learning from Setbacks

Through acceptance, you can view setbacks as opportunities for learning and growth, transforming challenges into stepping stones towards personal development.

(iii) Building Inner Strength

Acceptance cultivates inner strength by teaching you to confront difficult emotions and situations head-on, leading to a deeper sense of resilience and self-assurance.

When you approach life with acceptance, you pave the way for resilience and growth. By embracing acceptance and its transformative power, you forge a path towards a more resilient and empowered version of yourself.

6.1.7   Acceptance and Inner Clarity

Strengthening your capacity for exploration can lead to a deeper sense of inner clarity, illuminating the path towards understanding your true self and purpose. When you embrace exploration, you create space for inner peace to flourish within you. It's like a gentle breeze that clears the fog of confusion, allowing you to see things with a newfound clarity. This clarity isn't just about understanding the world around you; it's also about gaining insight into your own being.

Through exploration, you offer yourself the gift of self-compassion. You learn to treat yourself with kindness and understanding, acknowledging your flaws and imperfections without judgment. This self-compassion becomes a guiding light, nurturing your soul and helping you navigate life's twists and turns with grace.

As you probe deeper into exploration, you'll find that inner peace becomes a constant companion, calming the storms within you. This peace isn't just the absence of turmoil; it's a profound sense of harmony that emanates from embracing all aspects of yourself. In this state of inner clarity and peace, you can truly connect with your essence and uncover your purpose in this vast universe.

6.1.8   Transforming Through Acceptance

As you journey through life, accepting change can be a powerful catalyst for transformation. Embracing change with acceptance allows you to flow with the rhythm of life, opening up possibilities for growth and evolution. Finding peace through acceptance can lead to inner harmony and a deeper connection with your true self.

6.1.9   Embracing Change With Acceptance

Embracing change with acceptance allows for personal growth and transformation to unfold naturally. It's through this process that you can truly evolve and become the best version of yourself. Here are a few key points to ponder:

(i) Embracing Uncertainty

Accepting the unknown can be challenging, but it opens doors to new possibilities and experiences that can enrich your life.

(ii) Cultivating Trust

Trusting in the journey and in yourself is essential. It helps you navigate through changes with confidence and grace.

(iii) Letting Go of Resistance

Releasing resistance to change allows you to flow with life's currents, leading to a more peaceful and harmonious existence.

6.1.10   Finding Peace Through Acceptance

Roaming through the ebbs and flows of life's uncertainties can lead you to discover a profound sense of peace by embracing acceptance as a transformative force. Acceptance and healing are intertwined; they hold the key to finding inner peace amidst life's trials. When you accept the things you can't change, you free yourself from the burden of resistance. This liberation opens the door to a tranquil state of being, where you can find solace in the present moment. Healing begins when you let go of the need for things to be different and instead embrace them as they are. Through acceptance, you cultivate a peaceful sanctuary within yourself that allows for growth, understanding, and ultimately, profound healing.

6.1.11   The Path to Spiritual Growth by Acceptance

Beginning on the journey of spiritual growth can lead you to profound self-discovery and inner peace. It's a path that invites you to explore deep within yourself, uncovering layers of your being that you may not have investigated before. Along this transformative journey, you'll encounter challenges and victories, but each step will bring you closer to a heightened sense of self-awareness and personal growth.

Here below are the key points on the path of spiritual growth through acceptance:

(i) Exploration of Inner Truths

Engage in introspective practices such as meditation or journaling to connect with your innermost thoughts and emotions.

(ii) Embracing Vulnerability

Allow yourself to be vulnerable and open to the lessons that life presents, as vulnerability often leads to strength and resilience.

(iii) Cultivating Mindfulness

Practice being present in each moment, appreciating the beauty and lessons it holds, fostering a sense of gratitude and peace within.

6.1.12   Acceptance and Fulfillment

As you navigate the journey of spiritual growth, embracing acceptance opens the door to profound fulfillment within your being. Fulfillment arises from a place of deep contentment and peace, which can be cultivated through a dedicated mindfulness practice. When you accept things as they are, you free yourself from the burden of constantly reaching for a different reality. This acceptance doesn't imply resignation but rather a deep understanding that some things are beyond your control. By integrating acceptance into your spiritual growth journey, you build resilience to face life's challenges with grace and strength.

Mindfulness practice plays an important role in fostering acceptance and nurturing a sense of fulfillment. Through mindfulness, you learn to observe your thoughts and emotions without judgment, allowing you to accept them as they're without getting entangled in their grip. This practice enhances your spiritual growth by fostering a deep sense of inner peace and contentment, which in turn fortifies your resilience in the face of adversity. Embracing acceptance paves the way for profound fulfillment and resilience on your spiritual path.

6.1.13 Embracing Acceptance in Daily Life

In your daily life, consistently practicing acceptance allows you to cultivate a sense of peace and resilience in the face of life's challenges. Embracing acceptance is a powerful tool that can transform your outlook and bring about profound inner peace.

Here are three ways you can incorporate acceptance into your daily routine:

(i) Start Your Day with Daily Mindfulness

Begin each morning with a few moments of mindfulness. Focus on your breath, centre yourself, and set a positive intention for the day ahead. This practice can help you approach situations with clarity and calmness.

(ii) Practice Gratitude

Throughout the day, take time to appreciate the little things. Cultivating a sense of gratitude can shift your perspective towards positivity and abundance, fostering acceptance of what is.

(iii) Let Go of Control

Understand that not everything is within your control. Learn to let go of the need to micromanage every aspect of your life, and instead trust in the process of life unfolding.

6.2 About 'Acceptance' in the Bhagavad Gita

When we don't accept an undesired event it becomes anger,

When we accept it, it becomes tolerance.

When we don't accept uncertainty, it become fear,

when we accept it, it becomes adventure.

When we don't accept other's bad behavior towards us, it becomes hatred;

When we accept it, it becomes forgiveness.

When we don't accept other's success, it become jealousy;

When we accept it, it becomes inspiration.

6.3   Spiritual Surrender

When most people hear the word 'surrender', they think of something negative: giving up, losing, being humiliated, or allowing yourself to be controlled or perhaps even imprisoned. However, when used in a spiritual context, surrender also means to give up but here it's to give up everything that no longer serves you. Spiritual surrender is to stop struggling against "what is", let go of the smallness of life, and fully embrace its totality.

If there is one term describing an essential aspect needed for the success in life that can be found in all schools of spiritual practice, regardless of the methods prescribed and their scriptural foundation; one single aspect of approach or revelation of the Divine, the Self, the Eternal, the Perennial Ground of which all humans claim inheritance- it is ***surrender**".

You've tried everything and still aren't where you want to be. So stop struggling and let life move through you with spiritual surrender.

Surrender is giving oneself to the Divine—to give everything one is or has to the Divine and regard nothing as one's own, to obey only the Divine will and no other, to live for the Divine and not for the ego. Surrender is the decision taken to hand over the responsibility of your life to the Divine. By surrender we mean … a spontaneous self–giving, a giving of all your 'self' to the Divine, to a greater Consciousness of which you are a part.

There are many wrong ideas current about surrender. Most people seem to look upon surrender as an abdication of the personality; but that is a grievous error. … by taking the right attitude towards the Divine, this personality is purified of all the influences of the lower nature which diminish and distort it and it becomes more strongly personal, more itself, more complete. The truth and power of the personality come out with a more resplendent distinctness, its character is more precisely marked than it could possibly be when mixed with all the obscurity and ignorance, all the dirt and alloy of the lower nature. It undergoes a heightening and glorification, an aggrandizement of capacity, a realization of the maximum of its possibilities.

In the Vedic texts, spiritual surrender is referred to as the "joy of surrender." It is the wonderful, positive feeling you have when you simply let go. It's stepping out of all limitations, expanding beyond your usual conditioning, and opening to infinite possibilities. It's offering up the small self or personal identity to that of the absolute.

When you surrender spiritually, you stop forcing solutions on situations you can't control and instead trust and have faith that there is a Divine force taking care of everything in a perfectly orchestrated manner. As a practice, surrender is a way of unclenching your psychic and physical muscles. It is an antidote to the frustration that shows up whenever you try to control the uncontrollable. Here below are some ways in which surrender is further explained:

(i) In the Bhagavad Gita (18.62), Lord Krishna tells Arjuna:

tam eva śharaṇaṁ gachchha sarva-bhāvena bhārata
tat-prasādāt paraṁ śhāntiṁ sthānaṁ prāpsyasi śhāśhvatam.

Meaning: Surrender exclusively unto Him with your whole being, O Bharat. By His grace, you will attain perfect peace and the eternal abode.

(ii) Further in Bhagavad Gita (18.66), Lord Krishna, who represents the Divine, tells Arjuna:

Sarva-dharmān parityajya mām ekaṁ śharaṇaṁ vraja
ahaṁ tvāṁ sarva-pāpebhyo mokṣhayiṣhyāmi mā śhuchaḥ.

Meaning: Abandon all varieties of dharmas (purpose) and simply surrender unto Me alone. I shall liberate you from all sinful reactions; do not fear.

(iii) Spiritual teacher Eckhart Tolle says,

"Surrender is to say 'yes' to life—and see how life suddenly starts working for you rather than against you."

(iv) Author Debbie Ford says,

"Surrender is a gift that you can give yourself. It's an act of faith. It's saying that even though I can't see where this river is flowing, I trust it will take me in the right direction."

The ego loves to control everything. It creates the boundaries and limitations in which you imprison yourself. Surrender allows you to break free "the wisdom of uncertainty." With surrender, you can connect fully with your source where the whole universe conspires to support you. You can become timeless, eternal, infinite, joy-filled, and fearless. Individual aspirations and desires are surrendered to a higher power. Your individual desires dissolve into "Thy will be done."

6.4 Self-Surrender

Self-surrender is essential, and by that is meant the confession of personal impotence. I can of my own self do nothing. Since creation is finished, it is impossible to force anything into being. The example of magnetism is a good illustration. You cannot make magnetism; it can only be displayed. You cannot make the law of magnetism. If you want to build a magnet, you can do so only by conforming to the law of magnetism. In other words, you surrender yourself, or yield to the law. In like manner, when you use the faculty of assumption, you are conforming to a law just as real as the law governing magnetism. You can neither create nor change the law of assumption. It is in this respect that you are impotent. You can only yield or conform, and since all of your experiences are the result of your assumptions (consciously or unconsciously), the value of consciously using the power of assumption surely must be obvious. Willingly identify yourself with that which you most desire, knowing that it will find expression through you. Yield to the feeling of the wish fulfilled and be consumed as its victim, then rise as the prophet of the law of assumption.

6.5 Some Steps Towards Surrender

Now that you know why surrender is so powerful, the question becomes, how do you surrender? Here are **five steps** to follow on your journey toward surrendering.

6.5.1 Faith

To fully surrender, you must have faith, or as Indian philosopher Sri Aurobindo described it as "The soul's belief in the Divine's existence, its wisdom, power, love, and grace." No need to think of anything else, no need to go anywhere else, no need to deviate anywhere from your goal. It is a state of total confidence and respect for all things. This is to have faith in your spiritual teachers, in the teachings, and in yourself. The faith that helps you accept everything—the good and bad equally.

Once, the Mother (famous saint of Aurobindo Society) was asked 'What is the secret of success in sadhana?' to which she replied in one word 'Surrender'. In another talk the Mother says that surrender is the first attitude indispensable for beginning the Yoga. If one has not decided to make a total surrender, then one cannot even begin practicing Integral yoga.

6.5.2 Patience

You also need patience. Faith and patience complement each other. Each is both: the cause as well as the effect of the other. Both are the means as well as the end of the other. Between them they contain a complete code of conduct for a spiritual life. Indian spiritual master Shirdi Sai Baba emphasized them as necessary for harmony and well-being. The patience to allow things to unfold naturally with the understanding that everything you need will automatically come your way at the perfect moment.

6.5.3 Awareness

Surrender requires you to be aware. You need to remain focused and diligent, and overcome any doubts. While the normal concept of surrender is one of contraction, spiritual surrender is one of expansion. The more you surrender, the more your awareness expands and this can only come when you live with an alertness that stems from love.

If you have the slightest feeling that you are making a sacrifice, then it is no longer surrender. For it means that you reserve yourself or that you are trying to give, with grudging or with pain and effort, and have not the joy of the gift, perhaps not even the feeling that you are giving. When you do anything with the sense of a compression of your being, be sure that you are doing it in the wrong way. True surrender enlarges you, expands you; it increases your capacity; it gives you a greater measure in quality and in quantity which you could not have had by yourself. This new greater measure of quality and quantity is different from anything you could attain before: you enter into another world, into a wideness which you could not have entered if you did not surrender.

6.5.4   Meditation

Meditation is an act of surrender and the single most powerful tool you have on your spiritual journey. By turning your awareness away from the normal activity and settling to quieter and quieter levels of the mind, you can reconnect with your true, essential Self. When you slip into the silent spaces between thoughts, you surrender the small self and all its limitations, to your unbounded, eternal Self. You submerge your own identity into that of the Divine, where "I", "Me", and "Mine" disappear into the bliss of oneness.

As part of the daily meditations, you can ask yourself four questions:

- "Who am I?"
- "What do I want?"
- "What is my purpose?"
- "What am I grateful for?"

You can ask these questions at the level of your heart or your higher Self and then just listen to whatever answer arises from within, without evaluation. This too is a form of surrender. You are going beyond the ego to hear guidance from your higher being.

6.5.5 Bhakti Yoga

In the Vedic tradition's four paths to enlightenment, Bhakti is the path of love and devotion. This is also the path of surrender.

Patanjali, in the Yoga Sutra, famously describes the observance of Ishvara pranidhana—literally, surrender to the Lord—as the passport to samadhi, the inner state of oneness that he considers the goal of the yogic path. Among all the practices he recommends, this one, referred to casually in only two places in the Yoga Sutra, is presented as a kind of ultimate trump card. If you can fully surrender to the higher will, he seems to be saying, you basically don't have to do anything else, at least not in terms of mystical practice. You'll be there; however you define "there"—merged in the now, immersed in the light, in the zone, returned to oneness. At the very least, surrender brings a kind of peace that you don't find any other way..

Bhakti Yoga tells you to set aside time each day for the contemplation of the Divine to become attuned to it and its will; to recognize your life as spiritual and develop your finer levels of feeling. Serving humanity in everything you do and seeing everything as Yoga or union with the Divine.

In the case of bhakti - I approach Iswara and pray to be absorbed in Him. I then surrender myself in faith and by concentration. What remains afterwards? In place of the original 'I', perfect self-surrender leaves a residuum of God in which the 'I' is lost. This is the highest form of devotion (parabhakti), prapatti, or surrender You give up this and that of 'my' possessions. If you give up 'I' and 'Mine' instead, all are given up at a stroke. The very seed of possession is lost. Thus the evil is nipped in the bud or crushed in the germ itself. Dispassion (vairagya) must be very strong to do this. Eagerness to do it must be equal to that of a man kept under water trying to rise up to the surface for his life."

When we surrender to a strength and power of something greater than ourselves, our fear dissolves as we make room for a new host of possibilities and opportunities. We acknowledge that there may be a plan that far exceeds our own vision or understanding of our present circumstance or life. Of course, this doesn't absolve us of responsibility or the need to take action, but it does open the door for divine guidance so that we can take the next right action.

6.5.6   Through Prayer

To surrender to Higher Power and let go, you need strength and guidance as it's not always clear to understand God's plan for you. Through prayer, you can ask for strength to overcome your egoic desires, attachments, and beliefs which no longer serve you. You can also pray for guidance and clarity when your mind feels clouded during challenging times.

The secret to prayer is to forget what you think you need. Do not pray for a specific outcome. Ask the Universe to show you what you need. Ask for the highest good.

6.5.7   Be Mindful of your Thoughts

Most often our minds are clouded with worries, doubts and fears which make us feel like we are lost in our own darkness. Our fears and worries make it harder to surrender to Super power as we stay stuck in our comfort zone instead of choosing the path of personal and spiritual growth. Like every other discipline in life, we must practice mindfulness and learn how to become aware of our negative thoughts.

Practicing mindful meditation and breathwork exercises enhance self-awareness and help you become more mindful of your thoughts. During mindful meditation, you focus on the present moment instead of dwelling on the past or future.

6.5.8   Letting go of the Need to Control

Letting go of the need to control is one of the hardest things to do as we do not like feeling vulnerable. We are under the impression that if we control things, we avoid suffering and everything goes our way.

But in reality, we cannot control what is outside of us and thus we create more suffering. We can only control how we react to things, how we think and how we feel. By strengthening our faith in God, we will find it easier to let go and trust completely in our Creator.

When we surrender to God and let go, we feel relieved and have a sense of freedom. Even though it sounds scary, it's extremely liberating as it frees our minds from negative thinking, doubts and speculative fear.

6.5.9   Spend Time in Nature

Spending time in nature helps us feel more connected to God. It's extremely beneficial to go for a walk in the nature especially before or after a long day at work, or both in the morning and in the evening.

Contemplative walking, sitting down quietly in nature, watching the sunset or the sunrise etc…have a calming effect on our entire being. Psychologically we feel better, and even spiritually. By connecting to nature, you are connecting to God, our Creator.

6.5.10   Journal your journey

Journaling is a highly beneficial exercise that improves your quality of life. It helps you become in touch with your emotions and your thoughts while making connections with life circumstances and possible triggers.

Sometimes we feel overwhelmed with life and our own emotions. Writing down your story and how you are feeling helps you clear your head as you put everything on paper. By organizing your thoughts and releasing overwhelming feelings, you can experience moments of clarity and realization.

With a calmer mind and a clearer vision, it becomes easier for you to surrender to God, let go, and establish goals that are true to your authentic self. Journaling your story also helps you to achieve these goals once established.

Patanjali also said that devotion and surrender is the path to experiencing unending peace and entering higher states of consciousness. When you give everything to the Divine, you have nothing left to worry about.

"Love came and made me empty.
Love came and it filled me with the Beloved.
It became the blood in my veins.
It became my arms and legs.
It became everything. Now all I have is a name.
The rest belongs to the Beloved.

—Rumi

6.6   To Whom Do You Surrender?

This can be a difficult question for many people on a spiritual path. Do I surrender to a spiritual teacher or Guru? Am I submitting or selling myself out to some religious figurehead or deity that I don't even know? True surrender is never to a person, but always to the higher, deeper will, the life force itself.

No truly enlightened teacher will ever ask you to surrender your body, mind, or belongings to him or her. If you are fortunate enough to find an enlightened teacher, he or she will only accept what you freely choose to offer. When you become filled with the blessings of his or her pure unconditional love, it is easy to surrender.

Indian Yogi and Guru Paramhansa Yogananda said that spiritual surrender means "self-offering of one's will to God." It is not passive. It is an active surrender to the will of God. The manifestation of God will be different for different people based on their beliefs.

To offer everything to the Divine in whatever form you choose is the first and final step to letting go of the old and building the world that you've always dreamed possible. In fact, although in your possession, nothing really belongs to you, you just temporarily borrow everything.

However, when you truly surrender to any aspect of the Divine, human or otherwise, you are really surrendering to your own being in another disguise. Whatever you need on your spiritual journey isn't given, it's received. It's received by your ability to attune with your own higher Self and this ability develops with surrender, love, and devotion. Automatically you are elevated to your highest level so help doesn't come from outside but from your own being.

Complete surrender is the final fulfillment of your spiritual life. It isn't about giving anything up. You don't have to leave your family, quit your job, give away your possessions, or live in a monastery. Surrender isn't about doing,; it's about just allowing. Allowing the truth, beauty, purity, and goodness that is and always has been you, to flow effortlessly through you and radiate out into the world.

6.7 Significance of Surrender

A person in Krishna consciousness is a benefactor to all living entities and is peaceful and always **surrendered to Krishna**. He has no material desires. He is very humble and is fixed in his purpose. He is victorious over the six material qualities such as lust and anger, and he does not eat more than he needs. He is always sane and is respectful to others, but he does not require respect for himself. He is gentle, merciful, friendly, poetic, expert and silent.

In this regard, **Lord Chaitanya** also quotes a verse from *Kātyāyana-saṁhitā*: «One should rather tolerate the miseries of being locked in a cage filled with fire than associate with those who are not devotees of the Lord.» One is also advised not to even look at the faces of persons who are irreligious or who are devoid of devotion to the Supreme Lord. Lord Chaitanya recommends that one should scrupulously renounce the association of unwanted persons and completely take shelter of the Supreme Lord Krishna. The Lord is very kind to His devotees, and He is very grateful, able and magnanimous. It is our duty to believe His words, and if we are intelligent and educated enough, we will follow His instructions without hesitation. In *Śrīmad-Bhāgavatam* (10.48.26) Akrūra tells Krishna:

> *kaḥ paṇḍitas tvad aparaṁ śaraṇaṁ samīyād*
> *bhakta-priyād ṛta-giraḥ suhṛdaḥ kṛta-jñāt*
> *sarvān dadāti suhṛdo bhajato 'bhikāmān*
> *ātmānam apy upacayāpacayau na yasya"*

"Who can surrender to anyone other than Yourself? Who is as dear, truthful, friendly and grateful as You? You are so perfect and complete that even though You give Yourself to Your devotee, You are still full and perfect. You can satisfy all the desires of Your devotee and even deliver Yourself unto him."

A person who is intelligent and able to understand the philosophy of Krishna consciousness naturally gives up everything and takes to the shelter of Krishna. In this regard, Lord Chaitanya recites a verse spoken by Uddhava in *Śrīmad-Bhāgavatam* (3.2.23):

aho bakī yaṁ stana-kāla-kūṭaṁ
jighāṁsayāpāyayad apy asādhvī
lebhe gatiṁ dhātry-ucitāṁ tato 'nyaṁ
kaṁ vā dayāluṁ śaraṇaṁ vrajema

"How can one take shelter of anyone but Krishna ? He is so kind. Even though Bakāsura's sister planned to kill Krishna when He was an infant by applying poison to her breast and offering it to Krishna to suck and thus die, still that heinous woman received salvation and was elevated to the same platform as Krishna 's own mother." This verse refers to the time when Pūtanā planned to kill Krishna. Krishna accepted the poisonous breasts of that demonic woman, and when He sucked the milk from her, He sucked out her life also. Nonetheless Pūtanā was elevated to the same position as Krishna›s own mother.

There is no essential difference between a fully surrendered soul and a person in the renounced order of life. The only difference is that a fully surrendered soul is completely dependent upon Krishna. There are six basic guidelines for surrender. The first is that one should accept everything that is favorable for the discharge of devotional service, and one should be determined to accept the process. The second is that one should give up everything that is unfavorable to the discharge of devotional service, and one should be determined to give it all up. Thirdly, one should be convinced that only Krishna can protect him and should have full faith that the Lord will give that protection. An impersonalist thinks that his actual identity is in being one with Krishna, but a devotee does not destroy his identity in this way. He lives with full faith that Krishna will kindly protect him in all respects. Fourthly, a devotee should always accept Krishna as his maintainer. Those who are interested in the fruits of activities generally expect protection from the demigods, but a devotee of Krishna does not look to any demigod for protection. He is fully convinced that Krishna will protect him from all unfavorable circumstances. Fifth, a devotee is always conscious that his desires are not independent; unless Krishna fulfills them, they cannot be fulfilled. Lastly, one should always think of himself as the most fallen among souls so that Krishna will take care of him.

Such a surrendered soul should take shelter of a holy place like Vṛndāvana, Mathurā, Dwārakā, Māyāpur, etc., and should surrender himself unto the Lord, saying, "My Lord, from today I am Yours. You can protect me or kill me as You like." A pure devotee takes shelter of Krishna in such a way, and Krishna is so grateful that He accepts him and gives him all kinds of protection. This is confirmed in *Śrīmad-Bhāgavatam* (11.29.34) where it is said that:

martyo yadā tyakta-samasta-karmā
niveditātmā vicikīrṣito me
tadāmṛtatvaṁ pratipadyamāno
mayātma-bhūyāya ca kalpate vai

If a person who is about to die takes full shelter of the Supreme Lord and places himself fully under His care, he actually attains immortality and becomes eligible to associate with the Supreme Lord and enjoy transcendental bliss.

There are thirty-five items of devotional service, and they can be analyzed as follows: (1) hearing, (2) chanting, (3) remembering, (4) worshiping, (5) praying, (6) serving, (7) engaging as a servitor, (8) being friendly, (9) offering everything, (10) dancing before the Deity, (11) singing, (12) informing, (13) offering obeisances,

(14) standing up to show respect to the devotees, (15) following a devotee when he gets up to go to the door, (16) entering the temple of the Lord, (17) circumambulating the temple of the Lord, (18) reading prayers, (19) vibrating hymns, (20) performing *saṅkīrtana,* or congregational chanting, (21) smelling the incense and flowers offered to the Deity, (22) accepting *prasāda* (food offered to Krishna), (23) attending the *ārātrika* ceremony, (24) seeing the Deity, (25) offering palatable foodstuffs to the Lord, (26) meditating, (27) offering water to the *tulasī* tree, (28) offering respect to the Vaiṣṇavas or advanced devotees, (29) living in Mathurā or Vṛndāvana, (30) understanding *Śrīmad-Bhāgavatam,* (31) trying one›s utmost to attain Krishna, (32) expecting the mercy of Krishna, (33) performing ceremonial functions with the devotees of Krishna (34) **surrendering** in all respects, (35) observing different ceremonial functions.

To these thirty-five items, another four can be added: (1) marking one's body with sandalwood pulp to show that one is a Vaiṣhṇava, (2) painting one's body with the holy names of the Lord, (3) covering one's body with the remnants of the Deity covers, (4) accepting Krishna, the water which washes the Deity. These four additional items make thirty-nine items for devotional service in all, and out of all of these the following five are most important:

(1) to associate with devotees, (2) to chant the holy name of the Lord, (3) to hear *Śrīmad-Bhāgavatam,* (4) to live in a holy place such as Mathurā or Vṛndāvana, (5) to serve the Deity with great devotion.

These items are especially mentioned by **Rūpa Gosvāmī** in his book **Bhakti-rasāmṛta-sindhu.** The thirty-nine items above, plus these five items, total forty-four items. Add to these the twenty preliminary occupations and there are a total of sixty-four different items for conducting devotional service. One can adopt the sixty-four items with his body, mind and senses and thus gradually purify his devotional service. Some of the items are completely different, some are identical, and others appear to be mixed.

Śrīla Rūpa Gosvāmī has recommended that one live in the association of those who are of the same mentality; therefore it is necessary to form some association for Krishna consciousness and live together for the cultivation of knowledge of Krishna and devotional service. The most important item for living in that association is the mutual understanding of *Bhagavad-gītā* and *Śrīmad-Bhāgavatam*. When faith and devotion are developed, they become transformed into the worship of the Deity, chanting of the holy name and living in a holy place like Mathurā and Vṛndāvana.

The last five items-mentioned after the first thirty-nine-are very important and essential. If one can simply discharge these five items, he can be elevated to the highest perfectional stage, even if he does not execute them perfectly. One may be able to perform one item or many items, according to one's capacity, but it is the principal factor of complete attachment to devotional service that makes one advance on the path. There are many devotees in history who have attained perfection in devotional service simply by discharging the duties of one item, and there are many other devotees, like Mahārāja Ambarīṣa, who executed all the items. Some individual devotees who attained perfection in devotional service by executing only one item are: Mahārāja Parikshit, who was liberated and fully perfected simply by hearing; Śukadeva Gosvāmī, who became liberated and attained perfection in devotional service simply by chanting; Prahlāda Mahārāja, who attained perfection by remembering; Lakshmī, who attained perfection by serving the lotus feet of the Lord; King Pṛthu, who attained perfection simply by worshiping; Akrūra, who attained perfection simply by praying; Hanuman, who attained perfection simply by becoming the servant of Lord Rāma; Arjuna, who attained perfection simply by being a friend of Krishna's; and Bali Mahārāja who attained perfection simply by offering whatever he had in his possession. As far as Mahārāja Ambarīṣa is concerned, he actually performed all the items of devotional service. He first of all engaged his mind upon the lotus feet of Krishna. He engaged his words, his power of speaking, in describing the transcendental qualities of the Supreme Personality of Godhead. He engaged his hands in washing the temple of the Deity, his ears in hearing the words of Krishna and his eyes in beholding the Deity. He engaged his sense of touch by rendering service to the devotees, and he engaged his sense of smell by relishing the fragrance of the flowers offered to Krishna. He engaged his tongue in tasting the *tulasī* leaves offered to the lotus feet of Krishna, his legs in going to the temple of Krishna, and his head in offering obeisances to the Deity of Krishna. Because all his desires and ambitions were thus engaged in the devotional service of the Lord, Mahārāja Ambarīṣa is considered the leader in discharging devotional service in all kinds of ways.

Whoever engages in the devotional service of the Lord in full Krishna consciousness becomes freed of all debts to the sages, demigods and forefathers, to whom everyone is generally indebted. This is confirmed in *Śrīmad-Bhāgavatam* (11.5.41):

devarṣi-bhūtāpta-nṛṇāṁ pitṝṇāṁ
na kiṅkaro nāyam ṛṇī ca rājan
sarvātmanā yaḥ śaraṇaṁ śaraṇyaṁ
gato mukundaṁ parihṛtya kartam

"Whoever fully engages himself in the service of the Lord, O King, is no longer indebted to the demigods, the sages, other living entities, his relatives, the forefathers or to any man." Every man, just after his birth, immediately becomes indebted to so many people, and one is expected to discharge many kinds of ritualistic functions because of this indebtedness. However, if one is fully surrendered unto Kṛishna, there is no obligation. One becomes free from all debts.

Enjoy the practice of spiritual surrender. It will give you so much comfort and relief. You will be able to relax knowing that you are being taken care of. You'll begin to trust in a plan far greater than your own. The Universe will lead you to creative solutions and miraculous experiences. Let go and receive.

CHAPTER 7

FORGIVENESS

To forgive is to set a prisoner free and discover that the prisoner was you.

—*Lewis B. Smedes*

7.1   What Is Forgiveness?

Forgiveness can be defined as a freely made choice to give up revenge, resentment, or harsh judgments toward a person who caused a hurt, and to strive to respond with generosity, compassion, and kindness toward that person.

When we think of forgiveness, what comes to mind most often is about forgiving an offender. However, there are other forms of forgiveness as well:

- We may struggle to forgive ourselves.
- We may find ourselves needing to ask someone else for forgiveness.
- We may need to accept a request for forgiveness.
- We may find ourselves needing to find forgiveness related to existential concerns.

An example of this would be the need to forgive the world for the suffering that is present in it, or to forgive God for not preventing a death.

In forgiveness work, a person must come to recognize that suffering is not directed at us alone; rather, it is part of human experience. Forgiveness is a challenging area for most people, and confusion often exists about what it entails.

The following are important points to keep in mind about forgiveness:

- Forgiveness does not require us to reconcile with the offender and have continued contact. There are times when it is in our best interest to stay away from the offender.

- Forgiveness is a process that can take time; it is not just a decision we make quickly. To forgive generally requires some emotional and mental energy on our part.

- To forgive means that we have to fully accept what actually happened, how we were hurt, how our lives were affected by the offense, and even how we have changed as a result.

- When we do not forgive, we continue to give the offense and the offender power over us. To forgive is to become free to move forward.

- We need never forget what happened. In spite of our continued memory of the event, we nevertheless forgive and live life in the present.

- Forgiveness does not relieve offenders of their responsibility. If it is necessary to pursue justice, we can still take the action that is needed, such as pressing legal charges, filing complaints, or otherwise appropriately addressing concerns.

7.2 Choosing Forgiveness to Free Ourselves from Pain

No one can make this decision for us. We must be ready to reclaim the parts of our life that were affected. We may need assistance to do this, depending on such factors as who perpetrated the offense (oneself, a family member, a stranger, etc.) and the amount of suffering or loss involved. We might benefit from using a journal to write down our thoughts or to work through the process. Depending on the situation, it might also be helpful to talk with a trusted friend, a saint, a clergy member, or a therapist. Mobilize whatever resources will support you in this before continuing. If you find that you are becoming distressed, stop the activity and consider obtaining professional assistance.

7.3   Stages of Forgiveness

There are four stages of forgiveness:

(i) Uncovering

This stage is about gathering information about how the offense has hurt us, changed us, or cost us. Often this includes reflection on how it has preoccupied us mentally or emotionally.

(ii) Decision

Once we understand how not forgiving has cost us and what forgiveness is, we can decide to commit to the process.

(iii) Working

This stage is challenging. We work to gain a deeper understanding of the offender, our self, and the relationship, as applicable. During this stage, we begin to experience more empathy and compassion for our self and for the perpetrator.

(iv) Deepening

Finding meaning in the suffering might include becoming more connected to others or recognizing that suffering is universal.

7.4   Exploring a Grievance

- Are there events or situations in which you feel that you have been wronged and which continue to bother you?

- How did the event(s) change you? Did it change how you view your world, yourself, and others?

- What emotions are still present? Anger? Guilt? Shame? Hurt? Others?

- What has holding on to this cost you in terms of time, relationships, energy, happiness, etc.? Has it affected your health? Your sleep?

- How often does the event come to mind? How often do you dwell on it?

- What benefits might come to you if you were able to emotionally forgive the offense and/or the offender?

- Do you feel ready to do the work of forgiveness in order to free yourself from the past? Can you decide that forgiveness is for you and not the offender? If yes, continue with the following questions:

    (i)   What is left to express to the offender or about what happened to you? Write it down, express it through creative means such as drawing, or talk it out. You can also write a letter to the offender, outlining what is unexpressed. There is no need to send it out unless that is important to you.

    (ii) Have you needed forgiveness yourself from another person at some time? What was that experience like, and how did you feel? Recognize that everyone has been both the forgiven and the forgiver at some point.

    (iii) Is some of what you experienced through the offense impersonal (not really about you) but rather related to the suffering that is experienced by others on this planet?

(iv) If you put yourself in the other person's place, it can create empathy. You may ask, what led them to do what they did? This empathy, however, does not mean that their behavior is condoned.

- What can you learn from this experience? How can this connect you more with others? How can more of your energy go into what you value?

- Appreciate that the process may take time or require additional resources.

Forgiveness is an absolute necessity for continued human existence.

—*Bishop Desmond Tutu*

7.5   About Forgiveness in the Bhagavad Gita

tasmāt praṇamya praṇidhāya kāyaṁ, prasādaye tvām aham īsham īḍyam
piteva putrasya sakheva sakhyuḥ, priyaḥ priyāyārhasi deva soḍhum.(11.44)

Meaning: Therefore, O adorable Lord, bowing deeply and prostrating before You, I implore You for Your grace. As a father tolerates his son, a friend forgives his friend, and a lover pardons the beloved, please **forgive me** for my offences.

adveṣhṭā sarva-bhūtānāṁ maitraḥ karuṇa eva cha
nirmamo nirahankāraḥ sama-duḥkha-sukhaḥ kṣhamī (12.13)

santuṣhṭaḥ satatam yogī yatātmā dṛiḍha-niśhchayaḥ
mayy arpita-mano-buddhir yo mad-bhaktaḥ sa me priyaḥ
(12.14)

Meaning: Those devotees are very dear to Me, who are free from malice toward all living beings, who are friendly, and compassionate. They are free from attachment to possessions and egotism, equipoised in happiness and distress, and **ever-forgiving**. They are ever-content, steadily united with Me in devotion, self-controlled, of firm resolve, and dedicated to Me in mind and intellect.

śhrī-bhagavān uvācha

abhayaṁ sattva-sanśhuddhir jñāna-yoga-vyavasthitiḥ
dānaṁ damaśh cha yajñaśh cha svādhyāyas tapa ārjavam (16.1)

ahinsā satyam akrodhas tyāgaḥ śhāntir apaiśhunam
dayā bhūteṣhv aloluptvaṁ mārdavaṁ hrīr achāpalam (16.2)

tejaḥ kṣhamā dhṛitiḥ śhaucham adroho nāti-mānitā
bhavanti sampadaṁ daivīm abhijātasya bhārata (16.3)

Meaning (16.1-3): The Supreme Divine Personality said: O scion of Bharat, these are the saintly virtues of those endowed with a divine nature—fearlessness, purity of mind, steadfastness in spiritual knowledge, charity, control of the senses, sacrifice, study of the sacred books, austerity, and straightforwardness; non-violence, truthfulness, absence of anger, renunciation, peacefulness, restraint from fault-finding, compassion toward all living beings, absence of covetousness, gentleness, modesty, and lack of fickleness; vigor, **forgiveness**, fortitude, cleanliness, bearing enmity toward none, and absence of vanity.

CHAPTER 8

SPIRITUAL WISDOM

8.1   What is Spiritual Wisdom?

Spiritual wisdom can be defined as the understanding and knowledge that comes from a deep connection to one's inner self and the divine. Spiritual wisdom refers to a deep understanding of life's purpose, moral values, fostering personal growth, inner peace, and enlightenment. It is the ability to see beyond the material world and to understand the interconnectedness of all things. It is a realization that everything is interconnected and that there is a higher purpose to our existence. By seeking and cultivating spiritual wisdom, individuals can develop a greater sense of self-awareness, empathy, and compassion, leading to a more fulfilling and purposeful life. Spiritual wisdom goes beyond intellectual knowledge, focusing on personal growth and self-awareness. It often involves a deep connection to one's beliefs, values, and a higher power or universal force.

Challenges associated with cultivating spiritual wisdom can include overcoming ego, attachment, and resistance to change, as well as integrating spiritual wisdom into everyday life.

Spiritual wisdom transcends traditional education and intelligence, as it focuses on self-discovery and the understanding of our place within the universe. This cultivation of spiritual wisdom can lead to profound personal transformation and a greater sense of meaning in life, ultimately bringing us closer to a state of harmony and balance. By nurturing our spiritual wisdom, we not only deepen our connection to ourselves but also to others and the world around us.

One of the key principles of spiritual wisdom is the practice of mindfulness. This involves being fully present in the moment and being aware of one's thoughts, feelings, and actions. By being mindful, we can begin to understand the patterns and habits that may be holding us back and preventing us from living a fulfilling life.

Another important aspect of spiritual wisdom is the practice of self-reflection. This involves taking the time to reflect on one's thoughts, emotions, and actions in order to gain a deeper understanding of oneself. Through self-reflection, we can begin to identify and release limiting beliefs, negative patterns, and limiting emotions that are holding us back.

Spiritual wisdom also includes the practice of compassion and empathy. By understanding and accepting the humanity of others and ourselves, we can let go of judgement and criticism and develop a deeper sense of connection and understanding.

Developing spiritual wisdom enables individuals to live a more harmonious, fulfilling, and compassionate life, as they become more attuned to their inner selves and the world around them.

Spiritual wisdom involves developing a sense of gratitude and appreciation for the blessings in our lives. This can involve taking time to reflect on the things we are grateful for, or even keeping a gratitude journal.

Spiritual wisdom involves connecting with something greater than ourselves, whether it be a higher power, nature, or the universe. This can be done through meditation, prayer, or other spiritual practices.

Developing mindfulness, self-reflection, compassion, empathy, gratitude and connecting with something greater than ourselves are some of the key principles of spiritual wisdom.

Benefits of spiritual wisdom include a greater sense of inner peace, enhanced empathy and compassion, improved mental and emotional well-being, and a deeper connection to oneself and others.

8.2 About Spiritual Wisdom in Bhagavad Gita

shrī-bhagavān uvācha

abhayaṁ sattva-sanśhuddhir jñāna-yoga-vyavasthitiḥ
dānaṁ damaśh cha yajñaśh cha svādhyāyas tapa ārjavam (16.1)

ahinsā satyam akrodhas tyāgaḥ śhāntir apaiśhunam
dayā bhūteṣhv aloluptvaṁ mārdavaṁ hrīr achāpalam (16.2)

tejaḥ kṣhamā dhṛitiḥ śhaucham adroho nāti-mānitā
bhavanti sampadaṁ daivīm abhijātasya bhārata (16.3)

Meaning (16.1-3): The Supreme Divine Personality said: O scion of Bharat, these are the saintly virtues of those endowed with a divine nature—fearlessness, purity of mind, steadfastness in spiritual knowledge, charity, control of the senses, sacrifice, study of the sacred books, austerity, and straightforwardness; non-violence, truthfulness, absence of anger, renunciation, peacefulness, restraint from fault-finding, compassion toward all living beings, absence of covetousness, gentleness, modesty, and lack of fickleness; vigor, forgiveness, fortitude, cleanliness, bearing enmity toward none, and absence of vanity.

8.3 Spiritual Wisdom in Hindu and Buddhist Traditions

Different spiritual traditions emphasize various aspects of spiritual wisdom, such as enlightenment in Buddhism, gnosis in Gnosticism, and divine love in Sufism.

Both Hindu and Buddhist wisdom traditions offer pathways that promote personal development; personal responsibility, respect for all forms of life, preservation of peace, human welfare, nonviolence, altruism, integrity. While Hinduism is known for many themes like nonviolence and revering saints and teachers, "one of the central aims of this religion is the stability and welfare of the world". Buddha's teachings, the Dharma, shares these aspirational and practical learning outcomes. The ontologies of both traditions are based on perceptions that the essential nature of mind/consciousness is essentially pure with an intrinsic potential to awaken into that full realization. While the traditions are at variance regarding an essential self, atma in Hinduism, and anata, refutation of essential self in Buddhism, both traditions draw on the Law of Karma as a universal principle from which ethical reasoning and the development of **wisdom** is justified.

Karma

The law of karma is not regarded as rigid and mechanical, but rather a flexible, fluid and dynamic. Nevertheless there are relatively stable repeated patterns that arise from this collection of impersonal, ever-changing and conditioned events or processes, that form what we regard as a person's 'character'. To fully apprehend karma as truth depends on the cultivation of **wisdom**: cintamaya panna the **wisdom** obtained by thought, sutamaya panna the **wisdom** obtained by study, and bhavanamaya panna, the deep insight knowledge developed through meditation. In this way, for any person who recognizes dissatisfaction in themselves and seeks peace and happiness, openness to karma is heuristic.

Spiritual wisdom is a rare potential influenced by experience, age and cognitive, reflective and affective knowledge. It is also understood to be knowledge of what is good, personal standing of that good and a strategy to achieve that goodness.

For Tibetan Buddhist monks, wisdom includes attributes such as recognizing Buddhist truths, realizing emptiness is the true essence of reality; becoming the non-self; existing beyond suffering; being honest and humble; being compassionate to others; respecting others; treating all creatures as worthy and equal; having the ability to distinguish between good from evil; and being efficient in projects. Content analysis of the Bhagavad Gita, the most influential text of Hindu philosophy and religion, have found that for Hindu wisdom is related more to control over desires, renunciation of materialistic pleasures, emotional regulation, self-contentedness, compassion and sacrifice, insight and humility, yoga, decisiveness, duty and work love of God, and knowledge of life.

Lord Krishna advising his disciple Arjuna: On the battlefield of the Kurukshetra, Arjuna must decide whether to annihilate his own relatives. Krishna tells Arjuna that he should detach his inmost self, his eternal Atman or soul from his social role and then play that role without concern for personal consequences. True renunciation involves not renunciation of one's social role but renunciation of desires for the fruits of actions. The statement uncovers the **wisdom** features asserting, realization of truth of life's uncertainty, importance of responsibility towards society, control of desires and appropriate or skillful action. Several scholars successfully draw on the parallels between existential beliefs and values of the Bhagavad Gita teachings and modern psychotherapies and modern **wisdom** conceptualizations. Likewise, Jataka stories from the Buddhist canon, and well known throughout Asia, show how wisdom in education in plural classrooms can be furthered.

8.4   Measuring Wisdom

As interest in wisdom education grows, so does the quest to measure wisdom. Various wisdom measuring scales and methodologies are extant. One researcher reviewed and modified wisdom measurement tools and developed 53 Likert Scale items using the Delphi method. This study differentiates wisdom from intelligence on 46 items out of 49, however, 31 items differentiate spirituality from wisdom find symbiosis with Hindu and Buddhist wisdom domain. Commonly held constructs of wisdom include emotional regulation, practical life skills, life satisfaction, ethical conduct, self-esteem, mindfulness, and humility. Other studies throw light on the eastern characteristics and their consistency with universal wisdom components. Unanimously, prominent wisdom academics bring forth measurable and wisdom reinforcing capabilities of this rapidly increasing research discipline in the contemporary education. Another study draws on the role of different religious beliefs and their effects on the individual work values in organizations in dealing with daily work place issues. Interestingly, enough has been drawn on how wisdom can be measured but less on how some of the religious and spiritual values/beliefs contribute to holistic wisdom development and how these values/beliefs can be implicated as pedagogical tools in modern educational interventions.

CHAPTER 9

GRATITUDE

Gratitude, as it were, is the moral memory of mankind.

—Georg Simmel

9.1 What is Gratitude?

Gratitude is defined by the Random House Dictionary of the English Language as "the quality or feeling of being grateful or thankful".

Gratitude has been defined in a number of ways throughout history. Kant defined gratitude as "honoring a person because of a kindness he has done us". Scottish philosopher Thomas Brown defined gratitude as "that delightful emotion of love to him who has conferred a kindness on us, the very feeling of which is itself no small part of the benefit conferred". In psychological parlance, gratitude is the positive recognition of benefits received. Gratitude has been defined as an estimate of gain coupled with the judgment that someone else is responsible for that gain. Gratitude has been said to represent an attitude toward the giver, and an attitude toward the gift, a determination to use it well, to employ it imaginatively and inventively in accordance with the giver's intention. Gratitude is an emotion, the core of which is pleasant feelings about the benefit received. At the cornerstone of gratitude is the notion of undeserved merit. The grateful person recognizes that he or she did nothing to deserve the gift or benefit; it was freely bestowed. This core feature is reflected in one definition of gratitude as "the willingness to recognize the unearned increments of value in one's experience". The benefit, gift, or personal gain might be material or nonmaterial (e.g., emotional or spiritual). Gratitude is other-directed—its objects include persons, as well as nonhuman intentional agents (God, animals, the cosmos). It is important that gratitude has a positive valence: It feels good. It is described it as "intrinsically self-esteeming". Although a variety of life experiences can elicit feelings of gratitude, gratitude prototypically stems from the perception of a positive personal outcome, not necessarily deserved or earned, that is due to the actions of another person.

There are **three components of gratitude**:

(i)   a warm sense of appreciation for somebody or something,

(ii)   a sense of goodwill toward that person or thing, and

(iii)   a disposition to act that flows from appreciation and goodwill.

The virtue of gratitude is the willingness to recognize that one has been the beneficiary of someone's kindness, whether the emotional response is present or not. They thus conceived of it as a "moral virtue-trait" that leads a person to seek situations in which to express this appreciation and thankfulness. People feel grateful when they have received a benefit from someone who (the beneficiary believes) intended to benefit them.

9.2   Factors contributing to an Increased Focus on Gratitude

A number of contemporary trends have emerged that have helped for gratitude.

First, the positive psychology movement has directed attention toward human strengths and virtues—those inner traits and psychological processes that most cultures, philosophies, and religions have commended as qualities that fit people well for living in the world. Gratitude is a virtue, the possession of which enables a person to live well, and therefore must receive a hearing in any comprehensive treatment of the topic. The positive psychology movement has also called increasing attention to pleasant emotional states has referred to as the "sweetest emotions": happiness, joy, love, curiosity, hope, and gratitude.

Second, there is a renewed interest among social scientists in people's religious and spiritual lives. The roots of gratitude can be seen in many of the world's religious traditions. Thus, interest in personal manifestations of religion and spirituality may transport the scientist into the realm of gratitude. In the great monotheistic religions of the world, the concept of gratitude permeates texts, prayers, and teachings. The traditional doctrine of God portrays God as the ultimate giver. Upon recognition of God's outpourings of favor, humans respond appropriately with grateful affect, and gratitude is one of the most common emotions.

A **third** factor that makes this a propitious time for gratitude is the resurgent interest in virtue ethics, a subfield of moral philosophy. Philosophers have counted gratitude among the most important of the virtues, and as a necessary ingredient for the moral personality. Viewed through the lens of virtue ethics, gratitude is a purely person-to-person phenomena, apart from any reference to the divine. Ingratitude, on the other hand, is seen as a profound moral failure.

9.3   About Gratitude in Ancient Hindu Scriptures

As per Hindu scriptures (Taittiriya Samhita & Shatpath Brahman), we come into the world carrying debts we have to pay back over the course of our lifetime. And only when we pay this back are we able to achieve nirvana or moksha.

9.3.1   Dev Rin

The first is Dev Rin or debt to God: Think about it –what all do we have to be thankful to God for – our birth, our body, our family, our privileges, all of which came to us as good fortune and we took for granted. Isn't it? And the way to payback this debt is to be ever grateful for this in all thought and action.

9.3.2   Rishi Rin

The second is Rishi Rin) or debt to the saints/rishis who have come before us. Would we be able to do what we do if we did not have the knowledge created, discovered, articulated by intellectuals before our times?

9.3.3   Pitri Rin

The third is Pitri Rin or debt to our ancestors for the simple reason that they are the cause for our birth – a unique opportunity called life. This is also meant to induce in us a sense of trusteeship – we have got this earth – as a gift from our ancestors and we owe it to our next generation to hand it in a shape a gift should look like.

9.3.4   Manav Rin

The fourth is Manav Rin) or debt that we owe to all fellow human beings and the society at large. This debt can be repaid by treating others with respect and being of service to them.

9.3.5   Bhuta Rin

Finally comes the Bhuta Rin– Bhuta here is not the ghost you would think of but the pancha mahabhuta or the 5 elements – earth, water, fire, air and space. All of this material world is a combination of these 5 elements. Being indebted to them would mean indebted to everything; to plants, animals and nature at large.

9.4   About Gratitude in Buddhism

Gratitude is the quality of people of integrity, of good and noble character. Ingratitude, on the other hand, seems to be the rule of everyday life, when man displays his baser instincts. The Buddha, in the Suttas that follow, speaks about both and shows the strength of the former and the weakness of the latter.

The strength of the former is exalted in the Mailgala Sutta, the Sutta of Blessings, where the Buddha declares gratitude to be one of the highest blessings, thus showing how it plays a key role in His ethical and spiritual teachings. His message here is that if you cultivate gratitude, if you are grateful, this is a sign that you are making ethical and spiritual progress, that you are in the process of achieving a highest and rare blessing in life.

The typical weakness of ingratitude, on the other hand, is arrogance and egotism when thinking that whatever one has achieved in life was by one's efforts alone. A modest person realizes that the efforts of many people such as parents, teachers, friends, et al, have served in realizing his noble goals and he feels grateful to them. He sees the inter-connectedness of his existence. In contrast, an immodest person wishes "Let others think that this was achieved by me alone", and thus his desire and conceit do but only grow, to the point that he becomes utterly ungrateful, turning his back to his benefactors, even if and when they are in desperate need.

Normally it is difficult to know ungrateful people because they are devious.

"The ungrateful speak one thing with their mouth, Think another with their mind, Still do another with their body. This is the nature of vicious people. He who can know their nature, is certainly wise and knowledgeable."

Gratitude, however, is an uncanny and positive attitude of appreciation or thankfulness in acknowledging a benefit that one has received or will receive from others. It is often accompanied by a wish to thank them, or to reciprocate in kind, thinking: "This wasn't achieved by me alone, but by the help and support of this and that good person".

Psychological research has also suggested that feelings of gratitude may be beneficial to one's own emotional well-being as well. For example, Some researchers had participants test a number of different gratitude exercises, such as thinking about a living person for whom they were grateful, writing about someone for whom they were grateful, and writing a letter to deliver to someone for whom they were grateful. Participants [A] in the control condition were asked to describe their living room. Participants [B] who engaged in a gratitude exercise showed increases in their experiences of positive emotion immediately after the exercise, and this effect was strongest for participants who were asked to think about a [particular] person for whom they were grateful. Participants who had grateful personalities showed the greatest benefit from these gratitude exercises.

The conclusion was that people who tend to experience gratitude more frequently than do others also tend to be happier, more helpful and forgiving, and less depressed than their less grateful counterparts.

May you also by understanding the value of being grateful and cultivating it help you lead a happier and meaningful life.

9.4.1 Suttas about Gratitude

(i) Rare persons in the world

"There are two persons who are rare in the world. Which two? First, the one who volunteers to help others selflessly (pubbakari).

And second, the one who is grateful (katannu) and helps in return (katavedi)".

"The appearance of three persons is rare in the world. Which three? The appearance of a Buddha, ... The appearance of a person who can teach the Dhamma and Vinaya proclaimed by the Buddha, ... And the appearance of a person who is grateful (katannu) and helps in return (katavedi), is rare in the world."

(ii) Being Grateful is an Attitude of a Good Person

"A good person (sappurisa) is grateful (katannu) and helps in return (katavedi). This gratitude, this helping in return is praised by fine people. It is entirely the attitude of a good person."

(iii) Good Consequences of Being Grateful

"A wise person who is grateful (katannu) and helps in return (katavedi); who is a noble friend and has a firm faith in what is good; who attentively serves those in distress; such a one is called a good person (sappurisa). Prosperity does not leave him who is endowed with all these good qualities."

"Endowed with four things a good person is in heaven as though led and laid there. With what four? With bodily right conduct, with verbal right conduct, with mental right conduct and with gratitude and helping in return (katannutii, ka- taveditd)."

"By abandoning five vices, one can attain the first jhana, the second ... the fourth jhana; one can attain the sotapattiphala,... the sakadagami- ... the anagami-phala; one can attain Arahat-ship. Which five? Stinginess as to one's residence, stinginess as to one's supporters, stinginess as to one's gains, stinginess as to one's status, ingratitude and not helping in return (akatannutii, akataveditd)."

9.5 About Gratitude in the Ancient Philosophy

The importance of gratitude in ancient philosophical thought cannot be overstated. Many renowned philosophers from different traditions placed a strong emphasis on the practice of gratitude as a means to improve one's life, cultivate virtue, and achieve happiness.

Epicurus, an ancient **Greek philosopher** who founded the school of philosophy known as Epicureanism, believed that the key to happiness was in appreciating and being grateful for the simple pleasures of life. He advocated for a life of simplicity, focusing on the natural and essential desires such as friendship, freedom, and contemplation. By cultivating gratitude for these elements, individuals could lead a life filled with satisfaction and tranquility.

Stoic philosophers, such as Seneca and Epictetus, also placed great emphasis on gratitude. They believed that by practicing gratitude, one could develop a greater sense of contentment and resilience in the face of adversity. According to Stoic philosophy, gratitude allows individuals to better accept the events in their lives and maintain a sense of equanimity, regardless of their circumstances.

In ancient **Chinese philosophy**, Confucius emphasized the importance of gratitude in maintaining harmonious relationships and cultivating a virtuous character. He believed that gratitude was an essential component of the moral life and encouraged his followers to express their appreciation towards others and the natural world. Confucius also taught that gratitude should be directed not only towards others but also towards oneself, as self-appreciation is vital in maintaining inner balance and fostering personal growth.

9.6 Gratitude in World Religions

Gratitude is a universal concept that permeates various religious and spiritual traditions across the globe. By examining the role of gratitude in major world religions, we can gain a deeper understanding of its significance in shaping human values, beliefs, and practices.

In the **Christian faith**, gratitude is deeply rooted in the idea of thankfulness to God for his grace, mercy, and the many blessings he bestows upon his followers. The Bible is filled with references to giving thanks, with prayers, hymns, and psalms often expressing gratitude to God. Christians are encouraged to be grateful not only for the good things in life but also for the challenges and hardships, as these experiences can strengthen their faith and foster spiritual growth.

In **Islam**, gratitude is considered a fundamental aspect of faith and is closely connected to the concept of submission to the will of Allah. Muslims are encouraged to express gratitude for the countless blessings in their lives, from their health and well-being to the guidance and wisdom found in the Quran. The Islamic practice of daily prayers, known as Salah, is a way for Muslims to offer thanks and maintain a sense of humility and mindfulness of their dependence on Allah.

In **Judaism**, Gratitude plays a central role in religious practices and traditions. It is an essential aspect of daily prayers, blessings, and rituals. Jewish teachings emphasize the importance of acknowledging and appreciating the many blessings in life, both big and small. The practice of saying blessings before and after meals and observing the Sabbath are ways in which Jews express gratitude for the sustenance and rest provided by God.

Gratitude is an essential component of spirituality in **Hinduism**, as it fosters a sense of connection with the divine and helps cultivate a harmonious relationship with the world. In Hinduism, gratitude is expressed through prayers, rituals, and the practice of offering food or other items to deities. The principle of karma in Hinduism also highlights the importance of gratitude, as being grateful for one's circumstances can lead to positive actions and, ultimately, a more fulfilling life.

In **Buddhism**, gratitude is closely linked to the cultivation of mindfulness and the development of loving-kindness (metta) towards all living beings. Buddhists practice gratitude by recognizing the interconnectedness of life and expressing appreciation for the support and kindness of others. The practice of gratitude in Buddhism is seen as a means to counteract greed, attachment, and the illusion of self, ultimately leading to a deeper sense of compassion and interconnectedness with the world.

9.7   Science of Gratitude and Mental Health

Over the recent decades, the field of positive psychology has shed light on the numerous benefits of cultivating gratitude, with a growing body of research demonstrating its profound impact on mental health and well-being. By examining the science behind gratitude and mental health, we can gain valuable insights into how embracing thankfulness can lead to a happier, more fulfilling life.

Numerous studies have shown that individuals who regularly practice gratitude experience greater levels of happiness and overall life satisfaction. By focusing on the positive aspects of life and expressing appreciation for what one has, gratitude can shift the focus from negative thoughts and feelings to a more optimistic outlook. Gratitude has also been linked to a decrease in stress and anxiety levels. By acknowledging the good things in life and actively cultivating gratitude, individuals can better cope with stressors and challenges and become more resilient in the face of adversity. Gratitude has been shown to increase feelings of social support, which can be a protective factor against stress and anxiety.

Research has found that practicing gratitude can lead to better sleep quality and duration. Grateful individuals tend to have more positive thoughts before bedtime, which can help them fall asleep more easily and enjoy more restorative sleep. A good night's sleep, in turn, can lead to improved mood and overall well-being. Gratitude can also help individuals better manage their emotions and respond more adaptively to emotional experiences. By fostering a sense of appreciation for life's blessings, gratitude can help individuals maintain a balanced perspective and reduce the intensity of negative emotions such as anger, frustration, and sadness.

Studies have shown that gratitude can contribute to greater psychological resilience, enabling individuals to better cope with traumatic events and adversity. By cultivating a grateful mindset, people can develop a more robust mental and emotional foundation, making them better equipped to handle life's inevitable challenges. Gratitude can also improve interpersonal relationships by promoting feelings of warmth, empathy, and compassion. Expressing appreciation and thankfulness towards others can foster stronger connections, increase feelings of social support, and create more positive interactions.

In addition to the mental health benefits, gratitude has been linked to improved physical health. Grateful individuals tend to engage in healthier behaviors, such as regular exercise and maintaining a balanced diet, which can contribute to better overall health and well-being.

9.8 Finding Solace in Gratitude

Throughout history, gratitude has provided comfort and solace to people in even the most challenging circumstances. By recognizing the good in our lives and expressing thankfulness for it, we can foster a sense of perspective that allows us to find restfulness and contentment even in difficult times.

This practice helps us appreciate that, despite our struggles, there are always aspects of our lives for which we can be grateful.

By embracing gratitude, we can enhance our well-being, strengthen our connections with others, and find solace in even the most testing of circumstances.

CHAPTER 10

INTUITION

10.1   What is Intuition?

Intuition is that feeling in your gut when you instinctively know that something you are doing is right or wrong. Or it's that moment when you sense kindness, or fear, in another's face, you don't know why you feel that way; it's just a hunch. Intuition is not logical. It is not the result of a set of considered steps that can be shared or explained. Instead, while based on deep-seated knowledge, the process feels natural, almost instinctual. And yet, while intuition is quick and usually beneficial, it is not always entirely accurate. The subconscious brain attempts to recognize, process, and use patterns of thinking based on prior experience and a best guess. While intuition occurs in your day-to-day life, it is sometimes most apparent in the decisions of experts. The specialist draws on years of experience, held in unconscious frameworks, to make fast, high-quality decisions.

Intuition is soul guidance, appearing naturally in man during those instants when his mind is calm. The cultivation of intuitive calmness requires unfoldment of the inner life. When developed sufficiently, intuition brings immediate comprehension of truth. Intuition is a cognitive process, faster than we recognize and far different from the familiar step-by-step thinking we rely on so willingly. Unlike worry, the intuition will not waste your time. It is quick. It comes as a flash sometimes. God is the whisper in the temple of your conscience, and He is the light of intuition. Intuition provides us with a compass to navigate through life's complexities and reach alignment.

Intuition is a powerful survival instinct. An open mind, practice, and stillness can strengthen our ability to discern it. Although we often don't give it a second thought, intuition and spirituality are every bit as important as a doctor or friend in advising how best to prevent illness and heal ourselves. Intuition is our inner wisdom — the subtle nudges, gut feelings, and inner knowing that provide us with insights and guide us toward truth. Many spiritual traditions emphasize intuition as an important gateway for connecting to the divine within ourselves and the world around us. The more we tap into our intuition, the deeper our spiritual growth and understanding becomes.

Across cultures and history, developing intuition has been seen as a crucial component of spiritual awakening. Trusting our intuition allows us to tap into inspiration, creativity, and a deeper connection to ourselves and the world around us.

10.2   How Does Intuition Work?

What does psychology have to say about intuition, when much of what happens in the brain is invisible – like looking at the outside of a black box? Many psychologists propose a dual-process theory – decision-making processes split between intuitive (experiential or tacit) and analytical (rational or deliberate). Intuition (or blinking) typically refers to the use of knowledge that is not explicit. When it happens, it's hard to quantify or define, but it is there. The essence of intuition or intuitive responses is that they are reached with little apparent effort, and typically without conscious awareness. They involve little or no conscious deliberation. Intuition involves a sense of knowing without knowing how one knows. Intuitions also appear to be holistic – combining insights from multiple sources and often requiring a leap in thinking based on limited information. Intuition offers a reduction in overall cognitive load and the ability to respond instantly while providing confidence in our knowledge and decision making – even though it may defy analysis.

10.3   Processes involved in Intuition

People often make decisions – and reduce their cognitive load – based on what is good enough. Rather than arriving at complete and entirely correct answers, when faced with specific tasks, we often resort to heuristics – or rules of thumb – that help form intuitive judgments. The use of heuristics is considered commonplace and the default approach for making decisions. The process of recognition – a fundamental evolved function – is also crucial to intuition. It appears separate from other parts of the human memory in the brain, capable of persisting in the most challenging conditions with accuracy sufficient for practical purposes. Intuition appears to rely on the automation of the decision-making process.

10.4   Connection Between Intuition and Spirituality

Here are some of the key ways intuition intertwines with spirituality:

(i) Inner wisdom

Intuition allows us to tap into inner wisdom from our higher selves that goes beyond logic. Spiritual traditions believe this wisdom comes from a divine source.

(ii) Understanding Ourselves

Our intuition provides insights about our true nature, desires, and meaning. It helps reveal our unique spiritual paths.

(iii) Guidance

Intuition helps guide us towards our highest good and purpose. It provides support during challenging times and important decision-making.

(iv) Oneness

By trusting intuition, we deepen the mind-body-spirit connection and realize our oneness with all beings.

(v) Flow

Intuition allows us to detach from the ego and align our actions with divine will. We enter a state of effortless flow.

10.5 Spiritual Practices To Enhance Intuition

There are many spiritual disciplines that quiet the mind and create space for intuition to surface. Here are some of the top ones for boosting your innate intuitive abilities:

(i) Meditation

Meditation is highly effective for strengthening intuition, as it quiets mental chatter so you can tune into the subtle whispers of your intuition. Try setting the intention to connect with your inner wisdom before meditating. Pay attention to any intuitive nudges or inspirations that arise during or after your practice. Consistent meditation practice helps you detach from limiting beliefs and tap into intuition with more clarity.

(ii) Spending Time in Nature

Spending time in nature quickly quiets the mind by taking you out of the stimulation and information overload of daily life. Taking a walk in the nature, sitting under a tree, or viewing a sunrise/sunset are all ways to become more present and aware of intuitive insights.

Disconnecting from technology and artificial environments allows our intuitive capacities to heighten. Make an effort to spend more time in nature when you want to deepen your intuition.

(iii) Prayer and Journaling

Articulating your desires, questions and intentions through writing or speaking them activates your intuition. The simple act of bringing consciousness to what you want to understand ignites inner guidance.

Try journaling first thing in the morning to uncover intuitions brewing in your subconscious overnight. Or, vocalize prayers and then pay attention to intuitive nudges or absolute insights you receive.

(iv) Mindfulness

Mindfulness practices train you to become more present, clear out mental clutter, and perceive intuitive signs you may normally miss.

Whether you're washing dishes, walking to work, or having a conversation, do your best to stay fully present in the moment. This will heighten awareness and allow you to notice intuitive tugging.

(v) Visualization

Visualization involves clearly imagining desired outcomes, goals or situations. When combined with intuition, visualization allows you to perceive how to best achieve your desires.

Here's a simple visualization exercise: Close your eyes and imagine yourself already achieving a goal. Pay attention to any intuitive hits about concrete steps to manifest this outcome. Even 5–10 minutes of visualization can unlock powerful insights.

10.6　Tips For Strengthening Your Intuition

Here are some additional tips for awakening your intuition when combined with consistent spiritual practice:

(i) Set the intention

State a clear desire to strengthen your intuition. Setting this focus activates your subconscious mind to send intuitive insights your way.

(ii) Pay attention

Notice when intuition is flowing. For example, take note of gut feelings, spontaneous inspirations or answers that arise when you're contemplating an issue. Tracking intuitive moments trains you to tap into them more easily.

(iii) Keep an open mind

Avoid preconceived notions or rigid thinking, as they limit your ability to receive. Cultivate beginner's mind by staying open to all possibilities. This creates space for intuition.

(iv) Go with the flow

When you get an intuitive nudge, follow where it leads you even if it doesn't make total rational sense. This shows the universe you trust your intuition.

(v) Give yourself quiet time

Carve out periods during the day for silent contemplation without distractions. This allows subtle intuitive insights to come through.

10.7   Common Roadblocks in Intuition and Solutions

There are some common roadblocks that suppress intuition. Here's how to overcome them:

(i) Stress and Anxiety

When we're stressed or anxious, it's nearly impossible to perceive our intuition. Practicing calming spiritual activities like breathwork, yoga, being in nature, chanting or prayer can relax your nervous system so intuition can flow freely again.

(ii) Overthinking

Too much mental chatter drowns out our intuition. Regular meditation helps quiet the mind which allows us to differentiate between overthinking and true intuitive insights. Don't dismiss intuitions just because they seem too simple or irrational.

(iii) Technology Overload

Too much stimulation from devices, screens and information depletes intuition. Make an effort to unplug and spend time in nature or meditation to give your intuition a break.

(iv) Clutter

Physical clutter weakens our energetic vibration which decreases intuition. Maintaining a minimalist, orderly environment creates spaciousness mentally and energetically, allowing intuitions to surface powerfully.

(v) Skepticism

If you dismiss intuitions too readily, you will stop receiving them over time. Maintain an open, curious attitude about the guidance you receive. The more you act on your intuition, the stronger it will become.

Intuition is a gift available to us all. Integrating spiritual practices that quiet the mind makes space for our innate intuition to flourish. The more we pay attention and act on our intuitive insights, the deeper connection we experience with our truth and the divine.

Trusting our intuition allows us to tap into inspiration, transformation and purpose. We are able to align our actions with divine will and understand our souls more deeply.

Set the intention today to strengthen your spiritual intuition.

CHAPTER 11

VIRTUOUSNESS

11.1 What is Virtuousness?

Virtuousness is the quality of doing what is right and avoiding what is wrong.

A virtue (Latin: *virtus*) is a trait of excellence, including traits that may be moral, social, or intellectual. The cultivation and refinement of virtue is held to be the "good of humanity" and thus is valued as an end purpose of life or a foundational principle of being. In human practical ethics, a virtue is a disposition to choose actions that succeed in showing high moral standards: doing what is right and avoiding what is wrong in a given field of endeavor, even when doing so may be unnecessary from a utilitarian perspective. When someone takes pleasure in doing what is right, even when it is difficult or initially unpleasant, he/she can establish virtue as a habit. Such a person is said to be virtuous through having cultivated such a disposition.

The Greek term for virtue is *arête*, which was used for excellence of any kind. But generally the excellence referred to is an excellence belonging to human person so that the virtues may be described as the forms of human excellence. The opposite of virtue is vice. In ethics, 'virtue' is used with two somewhat different meanings. (a) A virtue is a quality of character – a disposition to do what is right in a particular direction, or to perform one of the more universal duties. (b) A virtue is also a habit of action corresponding to the quality of character or disposition. We may refer to the honesty of a human person, or to the honesty of his dealings equally as virtues.

Individual's values typically are largely, but not entirely, in agreement with their culture's values. Individual virtues can be grouped into one of **four categories** of values:

(i)   **Ethics** (virtue - vice, good - bad, moral - immoral - amoral, right - wrong, permissible - impermissible);

(ii)   **Aesthetics** (beautiful, ugly, unbalanced, pleasing)

(iii)   **Doctrinal** (political, ideological, religious or social beliefs and values), and

(iv)   **Innate/Inborn** (inborn values such as reproduction and survival).

11.2   Three Classes of Virtues

(i) Righteous Quality

There are virtues of what are known as, 'the righteous quality'. A virtue of this kind consists in the habit of performing a duty of a particular kind and in the quality of character which leads to this kind of action. The only distinction that can be made between virtuous conduct of this kind and right conduct is that the term 'virtuous conduct' emphasizes the habitual performance of what is right.

(ii) Requisite Quality

There are virtues of the 'requisite quality'. These are necessary to a virtuous character, but are also found in bad characters, and indeed may tend to increase the wickedness of the bad. Such virtues include prudence and perseverance. The villain who is persevering in his villainy is a worse man than the villain who is hesitant.

(iii) Generous Quality

There are virtues of the 'generous quality'. These are chiefly of an emotional kind and they add something not strictly definable, but of the nature of beauty or of moral intrinsic value, to actions that are in other respects right. They sometimes even give a strange quality of nobility to conduct that is morally wrong. We find this in the adventurous courage sometimes attributed to a brigand chief and in the loyalty of often shown to people utterly unworthy of that loyalty. Virtues of this kind seem to have some intrinsic value; this at least is suggested by the value that we assign to these virtues in the characters of people where no good result follows from the presence of the virtue in their actions.

Of the three classes, virtues of the '**righteous quality**' are the most important in the moral life. Virtues of the 'requisite quality' are clearly subordinate to the virtues of the 'righteous quality', for they are of value only when they accompany such virtues. Virtues of the 'generous quality' depend more on the natural endowments than the other two classes do, and are hardly to be acquired merely by the conscientious doing of one's duty. Virtues of this quality have an appeal that is perhaps more aesthetic than moral, but they do give to goodness a colour and an adventurous atmosphere which are sometimes sadly lacking in those whose virtues are merely of the righteous quality. Those who think of virtue as being something more than doing one's duty appear to be thinking often of some virtue of this kind, and these virtues do have about them a richness of emotion and a picturesqueness to which few people attain in the moral life.

11.3 About Virtues in Hinduism

Hinduism, or *Sanatana Dharma*, has pivotal virtues that everyone keeping the Dharma is asked to follow. For they are distinct qualities of *manushya* (humankind), that allow one to be in the mode of goodness.

There are **three modes of material nature (guna),** as described in the Vedas and other Indian Scriptures: (i) Sattva (goodness, creation, stillness, intelligence), (ii) Rajas (passion, maintenance, energy, activity), and (iii) Tamas (ignorance, restraint, inertia, destruction).

Every person harbors a mixture of these modes in varying degrees. A person in the mode of Sattva has that mode in prominence in one's nature, which one obtains by following the virtues of Dharma. The modes of Sattva are the following: Altruism: Selfless Service to all humanity;

Restraint and Moderation: This is having restraint and moderation in all things. Sexual relations, eating, and other pleasurable activities should be kept in moderation. Some orthodox followers also believe in sex only in marriage, and being chaste. It depends on the sect and belief-system, some people believe this means celibacy… While others believe in walking the golden path of moderation, i.e. not too far to the side of forceful control and total abandon of human pleasures, but also not too far to the side of total indulgence and total abandonment for moderation.

Honesty: One is required to be honest with oneself, honest to the family, friends, and all of humanity.

Cleanliness: Outer cleanliness is to be cultivated for good health and hygiene; inner cleanliness is cultivated through devotion to god, selflessness, non-violence and all the other virtues; which is maintained by refraining from intoxicants. Protection and reverence for the Earth.

Universality: Showing tolerance and respect for everyone, everything and the way of the Universe. Peace: One must cultivate a peaceful manner in order to benefit oneself and those around him.

Non-Violence/Ahimsa: This means not killing, or not being violent in any way to any life form or sentient being. This is why those who practice this Dharma are vegetarians because they see the slaughter of animals for the purpose of food as violent, when there are less violent ways to maintain a healthy diet.

Reverence for elders and teachers: This virtue is very important to learn respect and reverence for those who have wisdom and those who selflessly teach in love. The Guru or spiritual teacher is one of the highest principals in many Vedic based spiritualities, and is likened to that of God.

11.4   About Virtuousness in Upanishads

Katha Upanishad:

etacchrutvā samparigṛhya martyaḥ
pravṛhya dharmyamaṇumetamāpya.
sa modate modanīyaṁ hi labdhvā
vivṛtaṁ sadma naciketasam manye. . 1-II-13.

anyatra dharmādanyatrādharmā-
danyatrāsmātkṛtākṛtāt.
anyatra bhūtācca bhavyācca
yattatpaśyasi tadvada. . 1-II-14.

1-II-13. Having heard this and grasped it well, the mortal, separating the **virtuous** being (from the body etc.,) and attaining this subtle Self, rejoices having obtained that which causes joy. The abode (of Brahman), I think, is wide open unto Nachiketas.

1-II-14. Tell me of that which thou seest as distinct from **virtue**, distinct from vice, distinct from effect and cause, distinct from the past and the future.

Prashna Upanishad

pāyūpasthe 'pānaṃ cakṣuḥśrotre mukhanāsikābhyāṃ prāṇaḥ
svayaṃ
prātiṣṭhate madhye tu samānaḥ. eṣa hyetaddhutamannaṃ
samaṃ nayati
tasmādetāḥ saptārciṣo bhavanti. . III-5.

hṛdi hyeṣa ātmā. atraitadekaśataṃ nāḍīnāṃ tāsāṃ śataṃ
śatamekaikasyā dvāsaptatirdvāsaptatiḥ pratiśākhānāḍīsahasrāṇi
bhavantyāsu vyānaścarati. . III-6.

athaikayordhva udānaḥ puṇyena puṇyaṃ lokaṃ nayati pāpena
pāpamubhābhyāmeva manuṣyalokam. . III-7.

III-5-7: And then at the moment of death, through the nerve in the centre of the spine, the *Udana*, which is the fifth Prana, leads the **virtuous** man upward to higher birth, the sinful man downward to lower birth, and the man who is both **virtuous** and sinful to rebirth in the world of men.

Mundaka Upanishad

ehyehīti tamāhutayaḥ suvarcasaḥ
sūryasya raśmibhiryajamānaṃ vahanti.
priyāṃ vācamabhivadantyo 'rcayantya
eṣa vaḥ puṇyaḥ sukṛto brahmalokaḥ. . I-ii-6.

I-ii-6: Saying, 'Come, come', uttering pleasing words such as, 'This is your well-earned, **virtuous** path which leads to heaven', and offering him adoration, the scintillating oblations carry the sacrificer along the rays of the sun.

Brihadaranyaka Upanishad

sa vā ayamātmā brahma vijñānamayo manomayaḥ
prāṇamayaścakṣurmayaḥ śrotramayaḥ pṛthivīmaya āpomayo
vāyumaya ākāśamayastejomayo 'tejomayaḥ
kāmamayo 'kāmamayaḥ
krodhamayo 'krodhamayo dharmamayo 'dharmamayaḥ
sarvamayas
SF sa vai ayam ātmā brahma vijñānamayas manomayas
vāṅmayas k OM prāṇamayas cakṣurmayas
śrotramayas
ākāśamayas vāyumayas tejomayas āpomayas
pṛthivīmayas
krodhamayas akrodhamayas harṣamayas aharṣamayask
śrotramayas pṛthivīmayas āpomayas vāyumayas
ākāśamayas
tejomayas atejomayas kāmamayas akāmamayas
krodhamayas
akrodhamayas dharmamayas adharmamayas sarvamayaḥ
tadyadetadidammayo 'domaya iti yathākārī yathācārī tathā
bhavati.
tad yadā idammayas tad yad etad idammayas
adomayas iti yathākārī yathācārī tathā bhavati
sādhukārī sādhurbhavati
sādhukārī sādhus bhavati
pāpakārī pāpo bhavati
pāpakārī pāpas bhavati
puṇyaḥ puṇyena karmaṇā bhavati pāpaḥ pāpena.
SF puṇyas puṇyena karmaṇā bhavati pāpas pāpena
atho khalvāhuḥ
atha u khalu āhur
kāmamaya evāyaṃ puruṣa iti
kāmamayas eva ayam puruṣas iti
sa yathākāmo bhavati

sa yathākāmas bhavati
tatkraturbhavati
tathākratus tatkratus bhavati
yatkraturbhavati
yathākratus yatkratus bhavati
tatkarma kurute
tad karma kurute
yatkarma kurute
yad karma kurute
tadabhisampadyate. . IV-iv-5.

IV-iv-5: That self is indeed Brahman, as also identified with the intellect, the Manas and the vital force, with the eyes and ears, with earth, water, air and the ether, with fire, and what is other than fire, with desire and the absence of desire, with anger and the absence of anger, with righteousness and unrighteousness, with everything --identified, in fact, with this (what is perceived) and with that (what is inferred). As it does and acts, so it becomes; by doing good it becomes good, and by doing evil it becomes evil – it becomes virtuous through good acts and vicious through evil acts. Others, however, say, 'The self is identified with desire alone. What it desires, it resolves; what it resolves, it works out; and what it works out, it attains.

CHAPTER 12

SPIRITUAL ADAPTIBILITY

12.1   What does one mean by 'spiritual adaptibility'?

Spiritual adaptation or adaptibility is the power to adapt to spiritual changes or dangers. People's souls adapt to spiritual changes or dangers, as the soul can amidst changes to threats to their soul, circumventing potential irreparable damages to them or forcefully changing them to the desirable level. This can be considered as an evolving process in the spiritual sense, where this adaptation is the event of the soul adapting to the environment by making small changes depending on the environment, evolution resulting into the change of the soul by mutating at a stage due to adaptation processes. This can be a fallback defence mechanism or a guarantee for being accustomed for drastic changes to the persons' souls, allowing them to be as they are while also possibly expanding their metaphysical reach or power. Also, these persons can develop and use new and unique spiritual powers and techniques. Over the time, with the repetitive quintessential adaptations made unto the spirit, the soul becomes less and less of what it once was. Eventually growing into something else entirely over the course of the journey.

In the epic tale of Lord Rama's exile, we find profound leadership lessons that resonate even in our modern corporate landscape. This legendary narrative puts the spotlight on Lord Rama's unwavering resilience and remarkable adaptability as a leader amidst the trials and tribulations of life.

During his exile, Lord Rama faced an array of challenges that tested his mettle. Banished from his kingdom and separated from his beloved wife, Sita, Lord Rama navigated through the wilderness with a resolute spirit and a firm commitment to his principles. His journey was fraught with adversities, from encountering powerful demons to enduring the pain of separation. Yet, through it all, Lord Rama remained steadfast in his pursuit of righteousness.

In this compelling saga, we witness Lord Rama making critical decisions that showcased his unparalleled **adaptability**. He embraced uncertainty with grace, adapting to the ever-changing circumstances and learning from each experience. When faced with the enigmatic sage Agastya's advice, he incorporated newfound knowledge into his leadership approach, illustrating a leader's willingness to evolve and grow.

The significance of resilience and **adaptability** in modern corporate leadership cannot be overstated, especially in times of crisis and change. The business landscape is rife with uncertainties, disruptive technologies, and unforeseen challenges. Leaders who embody Lord Rama's unwavering resilience can weather storms with grace, inspiring confidence in their teams even in the face of adversity.

Adaptability, on the other hand, is the secret sauce that enables leaders to navigate uncharted waters successfully. Embracing change with an open mind and fostering a culture of **adaptability** empowers organizations to stay agile and innovative, driving growth and sustainability. Lord Rama's ability to **adapt** serves as a powerful reminder for leaders to embrace change as an opportunity for growth rather than a hindrance to success.

The saga of Lord Rama's exile urges us to reflect on our own leadership journey. How do we face challenges? Are we **adaptable** in the face of change? The timeless wisdom of this epic reminds us that true leadership is not solely about triumphs but also about learning, resilience, and the ability to **adapt** amidst life's uncertainties.

12.2 Laws of Leadership with Strong Spiritual Adaptability

Spiritual intelligence has been proposed as a measure of an individual's propensity toward spiritual leadership.

John C. Maxwell's seven Laws of Leadership provide an excellent framework for developing inclusive leaders with strong **spiritual intelligence**. Let's explore each of these laws and how they can be applied in the context of inclusive leadership.

(i) Law of the ability to connect

This law states that a leader's effectiveness is determined by their level of ability. Inclusive leaders with strong spiritual intelligence recognize that their ability to lead is not only based on their technical skills but also on their ability to connect with and inspire others. They work to develop their emotional intelligence and interpersonal skills, allowing them to communicate effectively, build trust, and create a sense of community.

(ii) Law of Influence

This law emphasizes that leadership is about influence, not just authority. Inclusive leaders with strong spiritual intelligence understand that they can't simply command respect from their followers. Instead, they focus on building relationships, connecting with others, and creating a shared sense of purpose. They inspire others through their words and actions, modeling the behavior they want to see in others.

(iii) The Law of Process

This law acknowledges that leadership development is an ongoing process, not a one-time event. Inclusive leaders with strong spiritual intelligence recognize that they are always learning and growing, and they prioritize their own personal development.

They seek out opportunities to learn from others, reflect on their experiences, and cultivate their inner wisdom.

(iv) Law of Navigation

This law emphasizes the importance of vision and strategy in leadership. Inclusive leaders with strong spiritual intelligence recognize that a clear sense of purpose and direction is essential for motivating and inspiring others. They take the time to craft a compelling vision and develop a strategy for achieving it, then communicate this vision clearly to their followers.

(v) Law of E.F. Hutton

This law states that when the real leader speaks, people listen. Inclusive leaders with strong spiritual intelligence understand the power of their words and use them carefully. They speak with intention, choosing their words to inspire and motivate others. They also listen carefully to others, seeking to understand their perspectives and needs.

(vi) Law of Solid Ground

This law emphasizes the importance of trust in leadership. Inclusive leaders with strong spiritual intelligence recognize that trust is the foundation of any effective relationship. They prioritize building trust with their followers, being transparent and honest in their communication, and following through on their commitments.

(vii) Law of Respect

This law states that people naturally follow leaders who are stronger than themselves. Inclusive leaders with strong spiritual intelligence recognize that strength comes not from authority or power, but from the ability to connect with and inspire others. They treat their followers and team members with respect and dignity, valuing their contributions and creating a sense of mutual trust and respect.

In summary, cultivating spiritual intelligence is critical for inclusive leadership development, as it allows leaders to connect with others on a deeper level, create a shared sense of purpose, and inspire and motivate their followers.

12.3   Typical Interview Questions connecting Spiritual Adaptability and Effective Leadership

1.   Tell me about your leadership experience.

2.   What do you consider are the important attributes of a leader?

3.   How would you describe a spiritual leader?

4.   What deeply motivates you?

5.   What guiding principles do you live by?

6.   What are your perceptions about big picture thinking?

7.   What are your perceptions about empathy?

8.   How do you feel about standing on your position even when it goes against mainstream thinking?

9.   What are your perceptions about adversity (in other words, when things go wrong how do you use what you learn from those situations)?

10.   Please share with me some words that sum up spiritual intelligence to you.

11.   Tell me your thoughts and ideas about trust.

12.   How do you perceive trust in the organization you work for?

13.   Share with me some words that demonstrate trust.

14.   What do you believe are the important attributes of a leader?

15. What are your perceptions of spiritual leadership in a business environment?

16. What is your perception of innovative, creative, and big idea ways of thinking in business?

17. What are your perceptions about empathy in the workplace?

18. What is your perception about business decisions that go against the popular proposals?

19. What do you believe is learned from situations where things go wrong in the workplace?

20. What is your perception of spiritual intelligence in a leader?

21. What is your perception about trust in your current organization?

12.4 Spiritual Adaptability as a Flexibility without Compromise

"You may bend me, but you will never break me."

–Roman Soldier Motto

We cannot be divided and conquered if our consciousness is high enough to transcend it. And so spiritual adaptation is important. Too much separation comes from the supposedly spiritual people, the religious people. They're actually not practicing the values and the teachings of their philosophy. They're not practicing it when they can't inter-associate with someone else who believes something different from them. "Nothing You Do Matters Unless What You Do Matters"

"Get Real or Die Trying"

We all come from different walks of life, different cultures (or lack thereof in some people's case), different religious and political backgrounds. But mostly we are often separated by consciousness.

More than all of these other, seemingly more obvious, factors. We like to think, "Oh, I grew up in the ghetto or it's race or it's religion. And this person grew up in the country. This person grew up in a wealthy family. This person was poor." These are all minor differences in comparison to consciousness differences, which we don't seem to put enough emphasis on in identifying things in interpersonal relationships with people. There are varying levels of consciousness and spiritual intelligence that connect us or separate us.

Without consciousness you separate. With consciousness, you connect. The more conscious you become, the less you actually judge, compete, and create separations from your fellow brothers and sisters, strangers, or anybody around you. The more spiritually aware and mature you become the more adaptable you do become. you can have strong beliefs and commitments yet be adaptable to the moment, environment, and most importantly to the soul in front of you. Now, more than ever, we are seeing division and separation; politically, religiously, and of course, racially. Isn't it interesting that those who claim to be the most devout and ascended and prescribe to the supposed spiritual beliefs and religious beliefs and dogma, whatever it may be, they're often the most judgmental and self-righteous people.

If we enter into conversations with a commitment to respect and connect with people, regardless of our disagreements, we are able to thoroughly enjoy ourselves, learn about each other, and even celebrate our uniqueness and differences. We are flexible and adaptable, to be fluid, and to be real. And while doing all this we do not compromise what we believed in. If we would have failed to do this, we would have been hypocrites and failed to live by our own ideals.

One has to be adaptable. Arrogant and prideful people are often rigid. Narcissists (overly self-involved people) literally can't comprehend someone else viewing something different from them. The world is just full of narcissists. And it's not always a common recognizable trait or identification that we label people as, but when you start studying the psychology, narcissism is very prevalent in human nature You can't change some things until you, until you start identifying it. But let's circle back to consciousness. When an individual is secure in their identity and their sense of self, purpose, and awareness, he doesn't really feel the need to project onto others, his feelings, and he certainly feels no need for others to believe and think the same things for others to validate them. The more spiritually conscious we become the less we need others around us to share our viewpoint. And that's an interesting point because it creates a type of spiritual independence. Now, of course, it is really wonderful to have kindred spirits around us. And in the long-term we all need people to be on the same page as us and believe what we do think, how we think, and share that.

The wisest and the most nourishing spiritual practices encourage us to experience our interconnectedness with others. We are individuals, but we are not islands. We are social beings as well as spiritual ones, inextricably connected in a network of relationships. That's why when we grow and change, we also experience uncomfortable growing pains in even our most long-standing relationships. Those on the spiritual path may recognize this all too well.

Our influence on each other is great - even when we're not consciously intending it to be this way. When we use our spiritual practice to grow and change the way we interact with the world and ourselves, we are also changing the usual role we've taken on in our systems. And as many wise seekers know, change is so often met with resistance.

Developing a spiritual practice means that you've probably assessed patterns and behaviors in your life to assess what serves you and your higher purpose and what should be left behind. While we all wish that the people in our lives will grant us this space to transform and be there to support us, it's just as common that we'll be met with skepticism, resistance, or at the very worst - resentment. This can be confusing, but when we lean on compassion and expand our perspective, we can understand why this happens.

Using systems theory as our guide, we see that systems have a balance to them. This balance is less of the harmony we feel between mind and body when we meditate or practice yoga and more of the dance of desires between our need to belong and our need to be an individual. Most of the time, when presented with change, we want to restore the balance as we knew it before. The familiar feels safe, even when it's not our highest path.

That's why people can feel fear in the face of growth - their own or another's. Your close loved ones may be faltering in their support not because they don't wish a fulfilling life for you, but because they are afraid of how their role may have to change in response to your newfound growth. You don't have to be on the spiritual path to know that change is scary, and we humans often try our hardest to resist it.

12.5   Spiritual Agility

Spiritual agility is a cluster of grit, emotional intelligence, and practice that allows us to respond to our changing realities with strength, speed, and stability. It is an important component of the spiritual intelligence.

So often, the human reality and the spiritual truth seem in conflict with each other. They seem to have different goals and focus. But we must build the bridge between the two. This is the path of Spiritual Agility. Spiritual Agility is the crux of everything.

In order to live a life that is truly fearless, we must bridge the gap between human reality and the spiritual truth. Our spiritual agility keeps us flexible as things change. In a moment, we could be experiencing joy, heart ache or fear, anything can happen at any time. But due to our spiritual agility, we are neither overjoyed by the favorable events, nor saddened by the unfavorable circumstances.

All the world's major religious figures and spiritual teachers at some point moved heroically with strength, speed, and agility to adapt their lives completely to a new framework of living. Buddha practiced spiritual agility when he ventured beyond the palace gate and saw suffering for the first time. It drove him into real, meaningful change that influenced millions beyond the banyan tree.

Leaders should be adaptable and open to innovative solutions. Staying flexible allows leaders to navigate changing circumstances and find new ways to succeed.

In the **Bhagavad Gita (Chapter 2, Verse 38)**, Krishna shares this timeless wisdom:

sukha-duḥkhe same kṛitvā lābhālābhau jayājayau
tato yuddhāya yujyasva naivaṁ pāpam avāpsyasi

Meaning: Fight for the sake of duty, treating alike happiness and distress, loss and gain, victory and defeat. Fulfilling your responsibility in this way, you will never incur sin.

CHAPTER 13

CONNECTING WITH HIGHER SELF

13.1 What is Higher Self?

The Higher Self is an extension of yourself that is part of you and part of the Source—the ultimate higher power, the universal source of cosmic energy. This energy links us to everyone and everything. We are all interconnected through shared energy. And as this universal energy—this higher consciousness—is contained within each one of us, we can use the Higher Self to access this universal wisdom and knowledge from within.

Once you know who you are and where you want to be, you'll likely find there is also a lot of work to be done to get to that place. This is a difficult but fulfilling process that can facilitate some of the most important and satisfying changes in your life.

In tandem with promoting spiritual connection, Higher Self Yoga also provides practitioners with practical tools, experiential exercises, and communication strategies from various schools of psychology to implement some of the changes they need to make in their life.

13.2 What is Lower Self?

The lower self is the self of passion and pride, lust and hatred, greed, resentment and ill-will, jealousy and envy. Many of us identify ourselves with this lower self of passions. It is also called the ego-self, which sits on the threshold of our consciousness. It easily catches us, captures our attention, and leads us astray on the path of evil and negativity. Trapped in its power, we are struggling in the darkness that is our own shadow.

13.3   Connecting to your Higher Self

Have you ever felt you are so much more than your outer activity and current life seems to indicate? Have you ever had the sense that inside of you somewhere is this great being of Light and power? That you had a deeper purpose? If so, then you already have an awareness of the wonderful being of Light that you truly are - your Higher Self! By reconnecting with your Higher Self, you will experience more joy, happiness, peace and abundance!

Finding meaning in your life is a key aspect of living well. Connecting with your higher Self is essential to experiencing your unique joyful life journey. Tapping into your higher Self allows you to embrace each moment fully and appreciate it for what it is.

Your higher Self is conscious. Much like how religion connects you with a higher entity, spirituality allows you to connect to your own conscious guide - your higher Self.

As a high-achieving person, you tend to move through life at a fast pace. You are often motivated by the desire to reach certain goals. This can result in feelings of urgency to finish one thing and move on to the next. And when your mindset is that you'll be happier and more successful once you reach that goal, it's easy to find that you're sacrificing your health, wellbeing, relationships, or dreams.

When you get in touch with your higher Self through spiritual practice, it's the journey that brings you happiness and success rather than the achievement of the goal itself. By connecting with your Self and discovering your own unique journey, you are better able to experience success without sacrifice.

Life's joyful journey starts with a conscious connection with your higher Self. Spiritual practices are how you strengthen this connection.

13.4 How to connect with your Higher Self?

(i) Connect with your lower self, first

Before you can delve into such a sacred undertaking as connecting with your higher self, you must connect with your lower self. We all share the same lower self, and that is Mother Earth. You cannot reach Her via meditation but rather by being joyous, playful and truly becoming a kid again. You will know when you have made the connection because you will never feel safer or more exuberant than when that time comes. If this first step isn't completed, the journey to connecting with the higher self becomes virtually impossible.

(ii) Nature and Sunlight

As simple as the concept of spending time outside may seem on the surface, there are ways we can use nature's ancient wisdom to further our own spiritual journeys. There have been many scientific studies throughout the years that have proven how connected we are with nature and the sun. These positive feelings from nature's vibrations allow us to connect with our hearts and ultimately awaken the Higher Self.

"Take half an hour to sit in the sun. Cover your head and always wear some suntan lotion. As best you can, feel the sun's warmth coming into your body. Experience it flowing into your heart, through all your limbs, through your stomach, etc. Really concentrate on the sun's energy permeating your body. Sun energy is very calming and relaxing. When you experience it, do so with an open heart and mind, and try not to think about anything; simply experience the warmth of the sun. You aren't meditating, but just staying with the warmth of the energy and the feeling of your body. This can be very healing, especially if you are tired and under any stress.

(iii) Meditation

When your mind is overactive, it can feel like your thoughts are going in endless circles. The busier the mind is, the harder it is to focus, relax, and most importantly, connect with our Higher Selves. The purpose of meditation is to quiet the mind so that we may turn our attention inwards. Meditation is the most important practice to uncovering and assimilating our true inner selves. It is the key to understanding and communicating with our higher power, inner wisdom, and our Higher Selves, which holds the promise of inner and outer personal change.

There are many different forms of meditation, but at its core, it simply requires sitting down in a comfortable place away from everyday distractions and quieting our busy minds for a short period of time. This can be done in a comfortable chair; there is no need to sit cross-legged on the floor while making an "ok" sign with your hands.

(iv) Find Your Vocation

Choosing the right career (your "vocation" or a "true calling") may not seem like a spiritual choice, but it's actually heavily tied into the Buddhist and Hindu ideas of dharma. One of the many meanings of dharma refers to the work a person chooses to do in a given lifetime. This work relates to the individual's evolving soul, and the work we do can help our souls go onward to another level of maturity. This does not mean that we have to achieve fame or amass huge amounts of wealth, as those are not true metrics of fulfillment. Instead, it's only important that our souls feel they have done the best work possible.

Since most of us spend over a third of our waking hours at our jobs, it's essential that we choose a vocation that's aligned with our spiritual path. Those of us with a deep desire to help others can advance their progress with a career in medicine or social work, the creatively-inclined can often find a fulfilling career in the arts, and even the mathematically-minded among us can achieve spiritual fulfillment with a career in engineering or accounting. As long as you listen to your Higher Self when it directs you to your soul's true vocation, you'll find yourself nurturing a deeper connection to your Higher Self every day of your career.

(v) Strengthen Your Intuition

Do you sometimes feel like "your gut" is trying to tell you something that you can't quite put your finger on? This subtle but persistent feeling is your intuition, and it's the reason you sometimes feel drawn toward a decision without fully understanding why. In a society that largely values only what can be rationally or empirically explained, it's no surprise that many of us ignore or repress the rather abstract sixth sense of intuition.

It's not a negative reflection of us; it's simply how most of us have been raised in society. But as a result, we may sometimes feel as though we're only working with a piece of our power instead of the full scope of hidden knowledge available to us. Learning to listen to our intuition is a key step of your spiritual development, and can further strengthen your connection to your Higher Self.

(vi) Work With Your Heart

Wisdom traditions tell us that there is some knowledge that cannot be found through intelligence, but through surrender and experience. It is only through surrender to the heart that we discover our higher knowledge.

Listening to your heart's wisdom is essential to progressing in your spiritual journey and developing a deeper connection with the Higher Self. When the heart is open, we become capable of the compassion, kindness, openness, and honesty we long to have shown back to us in our daily lives.

(vii) Cultivate Positive Speech and Thoughts

As part of our journey to connect to our Higher Selves, it's important that we remain conscious of what we say and think. It is within our lower nature to sometimes gravitate towards negative speech and thoughts, and the first step in making a change towards more positivity is to foster a heightened sense of awareness of what we say and think.

With time and practice, we can reach a place where more positive speech becomes your second nature. Working on our speech and thought patterns will simultaneously improve our patience (a skill that we often lack), especially when it comes to promoting positive change within ourselves.

(viii) Self-Reflection

Whatever paths you take to deepen your relationship with your Higher Self, you won't receive the full benefit of your training without regular and honest self-reflection. It's essential that we regularly take a step back, and think about the progress we've made and where we can improve.

Much like meditation, this is best achieved in a state of quiet, relaxed contemplation. The goal here is not to be self-critical or harsh, but simply to understand how your journey is progressing. Ask yourself what practices have been the most helpful, and how they've helped move you forward on your path. By the same token, consider which practices have been the least helpful. Is it because of the way you're going about them, or do you simply need to replace them with different practices? Is your personal life outside of your spiritual training affecting your practice? Are there changes you need to make in your life, your career, or your friendships?

(ix) Forgive yourself and others

Do not hold on to the past. Realize that you don't have to "forgive and forget". You can still acknowledge that you or someone else did something "wrong" while at the same time forgiving. Non-forgiveness binds you energetically to the other person and your lower self/ego.

(x) Journaling

Letting your thoughts flow from your body onto paper can help you connect with your conscious mind. The more often you write about your thoughts, feelings, and observations, the more you access your higher consciousness. Be consistent and you'll tap into your higher Self and what makes your life journey joyful.

As you deepen your connection with your higher Self, notice the shifts in your life. Embrace these changes, as they can be signs that you are on the path of your joyful journey and experiencing success without sacrifice.

(xi) Identify and get rid of limiting beliefs

We all pick up limiting, false beliefs along the way which keep us from being all we can be. Those beliefs can even cause pain and suffering. Resolve to really look at the things you believe in and WHY you have the belief. Realize that no one is forcing you to believe anything and that beliefs are NOT a part of who you really are! Determine to keep an open mind and know that limiting beliefs are keeping you from fully connecting with your True Self. Clearly see that false beliefs are restricting you from experiencing all the goodness of life, as well as contributing to suffering and a sense of struggle in life. Determine to let them go!

(xii) Reading

While nurturing a relationship with your Higher Self often revolves around learning new exercises and techniques, no spiritual practice is complete without traditional reading (like you're doing now!). Check out ancient books on Spirituality, for example, Vedas, Upanishads, Bhagavad Gita, Ramayana, Puranas, Manusmriti, etc., in Hinduism.

(xiii) Fill every day with gratitude

As you go through your day, regularly express your gratefulness for everything you can think of - the wonderful day, the beautiful flowers, the sunshine, your spouse, the smile on your child's face, your family's health, and on and on. Let gratitude fill your heart and you'll be surprised to see what a difference this will make! Gratefulness opens the doorway to your True Self.

(xiv) Maintain Harmony

Harmony is one of the major keys - if not THE major key - on the path to raising your consciousness and reuniting with your Higher Self. Harmony is an internal quality springing from the heart and not just an external expression. Harmony involves mastering your emotions and not letting them control you. State your clear intention to be centered in Harmony and to be harmonious in every situation you meet. Keep centered in the heart as much as possible and focus on peace and harmony.

(xv) Learn to stay neutral

Neutral is a powerful state of emotional balance and inner quiet and peace. It is a place of equilibrium where you are more able to access the wisdom of your Higher Self. Learning to stay neutral is a key to becoming non-reactive to situations. Being neutral is not a state of not feeling anything, but rather a state where you are in balance with your emotional responses to what is happening - neither reacting negatively or becoming overly excited.

(xvi) Coaching through a Guru

Working with the Higher Self takes time, commitment, and often, the encouragement and wisdom of others. A personal coach or a Guru is an important part of your spiritual development. A coach provides valuable insights that facilitate growth and a deeper relationship with your Higher Self — for beginners and experienced yogis alike. No matter where you are in your journey, coaching is critical for learning, growth, and self-awareness.

13.5   What to expect on Connecting with Higher Self

(i) Letting go

As you connect on your journey, you'll notice things that no longer serve you. Habits you've picked up over the years, how you use your time, and even some relationships may need to be reevaluated. Letting go of what no longer serves you can be a challenge. Yet it's part of the growth process, and releasing these things creates space that can be used in a more meaningful way as you continue your journey.

(ii) Core values

Your core values become clear and present once you connect with your Higher Self. You learn how to effortlessly incorporate them into your life. When your path yields success without sacrifice, your core values have become an integral part of your life.

(iii) Manifestation

Create a vision and hold on to it. When you're connecting with your higher Self, your vision is clearer and manifesting occurs with greater ease. Tapping into that clear vision leads you to your unique joyful journey.

Connect with your higher Self. Begin a journey that brings you purpose, fulfillment, and joy with greater ease. A joyful journey finds happiness in each moment, not merely when you achieve a goal.

CHAPTER 14

CONNECTING WITH THE UNIVERSE

Have you ever felt like the Universe has a secret language, speaking to us through the intricate web of life's lessons? It's like an invisible force nudging us toward personal growth and transformation.

Many persons have come to the realization at some point in their lives that there is a greater meaning, or purpose, to who they are and this life they are living. Understanding that there is a bigger meaning to life is only the first step to connecting with the Universe. The Universe is vast and complex and in our small space on this earth we want to know the best time to connect with the Universe, what does it mean to be connected to the Universe and even signs to know we are connected to the Universe.

14.1 Reasons to Connect with the Universe

(i) Mother Earth carries a healing, nurturing and comforting energy and when you choose to connect with her energy you will feel supported each step of the way on your spiritual journey.

(ii) We are cyclical beings, meaning we move through cycles of lessons that encourage us to grow and evolve on our spiritual journey. When we resist the cycle of learning we stay stagnant in one cycle, never evolving and continuously repeating the same challenges. By connecting to the Universe spiritually we open ourselves to the cyclical energy that surrounds us and we understand our own cycles from a higher perspective.

(iii) The Spring Equinox* carries the energy of the beginning of a new cycle. This Universal Energy amplifies our own energy and if we resist this energy and the lessons that come with it then we end up repeating the cycle we've already lived through which perpetuates the idea that our past dictates our future. When we choose to connect with the Universe spiritually through the Spring Equinox energy we prove that energy is limitless and always changing and evolving. Our past does not need to define our future as long as we are willing to learn the lessons and up-level into the next cycle.

*(During the equinox, the Earth's axis and its orbit line up so that both hemispheres get an equal amount of sunlight. The word equinox comes from two Latin words meaning equal and night).

(iv) Growth occurs over time, through a cycle, and we can see this lesson through Mother Earth's energy as she evolves through her own cycle. In spring seeds are planted as the soil awakens from a cold winter with the purifying energy of the rain. In the summer the extended days bring more sunlight so the plants can gain more energy to grow. In the fall we are able to harvest the blessings from the summer's growth. In the winter the earth needs to rest and heal from a cycle of growing in order to prepare for the next cycle. This energy is replicated into our own lives, as we can see how we grow as children into adults. We can see when we work on a project it's not completed in one day because there are a lot of steps to reach completion. Throughout your growth challenges may occur and we need to take the time to overcome these challenges which slows the growth period. When we connect to the Universe spiritually through the growth cycle of Mother Earth we can feel supported each step of the way on our own growth journey.

(v)   When you are feeling stuck in a challenge and unable to see the solution, you can connect with the Universe spiritually and ask your Angels and Spirit Guides for clarity. When you are connected to the Spirit of Mother Earth and the Animal Realm you are able to recognize the signs and messages from your Angels in your physical reality. If your Spirit Guides send a Deer on your path, the clarity would be centered around bringing compassion into the situation in order to overcome the challenge.

(vi)   The energy of the Universe is always reminding you that you aren't alone. When you slow down and connect to the Universe you can feel the energy of support and community fill you up.

(vii)   The best time to connect with the Universe is whenever you want clarity, need courage to step forward, desire support & comfort, need to be filled with purpose, feel a shift in your own energy, or you know there is a shift in energy in your environment. Basically anytime is the best time to connect with the Universe. If you are wanting to amplify specific energies within you then there is always a better time to connect with the Universe. For instance if you are ready to create a new beginning for yourself or call-in fresh energy, then the Spring Equinox is the best time to connect with the Universe spiritually.

(viii) In order to heal your past pain, the Winter Season is the best time to connect with the Universe spiritually. But, this does not mean it's the ONLY time to heal your past pain. When the Moon Cycle passes through the Full Moon phase and starts to decrease in size through the waning phases the energy that is amplified is reflection, releasing and healing, so you can slow down and take the time to connect with the Universe through the Moon's energy in order to feel supported as you heal your past pain.

(ix)   When you are living in survival mode your ego is easily triggered by your external environment. This energy leads you into the worry, doubt, fear and pain of believing you aren't safe. By connecting with the Universe spiritually you are able to find strength and courage in order to heal your shadows and past trauma. By connecting with the Universe spiritually you are able to alchemize the heavy energy of outdated patterns and unsupportive conditioning from society and generational trauma. By connecting with the Universe spiritually you are able to harmonize your shadow and light and allow abundance to flow into your life. Step by step you heal the parts of you that are clinging to survival mode and you shift your entire state of being into creation mode. This is the power of connecting with the Universe.

(x)   To be connected to the Universe means that you recognize that you have access to limitless energy and there is limitless support surrounding you always. Do you need more strength? You can connect to the Universe through the energy of mountains. Do you need more inspiration? You can connect to the Universe through the radiant energy of the sun. Do you need more trust? You can connect to the Universe through the energy of the Eagle. You have access to all of the abundant energy; you simply have to acknowledge the Universe and choose to accept the uplifting energy it's sharing with you.

(xi)   When you receive signs and messages that feel uplifting, supportive, encouraging, and you are instantly reminded that you aren't alone in this spiritual journey you will know that you are connected with the Universe. If an animal pops up on your path and immediately you feel it is energy filling you up and appreciation flows through your mind and heart, you will know you are connected with the Universe. When a challenge occurs on your path and you remember to ask your Angels and Spirit Guides for assistance and a calmness settles in, you will know you are connected with the Universe.

14.2   Steps to connect with the Universe

Connecting with the Universe begins with finding a quiet space to be alone with your thoughts. Sit in a silent room, close your eyes, and take a deep breath. As you delve into the depths of your being, ask a simple question, "Are you there?" Be patient and attentive, for the Universe's response may come in various forms — a whisper, a feeling, or a fortunate sign. Feel the love and energy surrounding you, and respond with gratitude, saying, "I love you."

The cosmos that consist of galaxies, stars, planets, space, and all other forms of matter and energy is called the Universe. As all matter is energy, it means that the Universe is simply an abundance of energy.

According to quantum physics, a quantum is an indivisible unit of something and everything is made up of quantized fields. The fundamental blocks of all matter and energy, that is the atoms, protons, neutrons, electrons, and photons, etc. are not particles but bundles of field energy.

The entire Universe is made up of different types of quantized fields that are unified into a single quantum field. Therefore, the Universe, including the earth, is nothing but a quantum field. In a quantum field, reality can exist in infinite possible states. Thus, the Universe is a field of infinite possibilities.

Since the Universe is a field of abundant energy that has infinite potential and possibilities, being a part of it, we also inherit this power of the Universe.

But the problem is that we do not remember this inheritance and also the fact that we're already connected to the Universe by default. We think of ourselves as separate from the Universe or the Quantum Field. Our ego (the sense of "I") has created the sense of individuality that limits us and makes us believe that we're a separate entity apart from the Quantum Field.

However, if we overcome this obstacle, by remembering our connection to the Universe, we can realize our forgotten potential, and then all we need to create the life we want is just tap into this abundant resource. Beyond the material needs, our health and well-being including the sense of peace, love, and joy can become everyday experiences.

We can manifest the life we want if we align with and tune in with the frequency and vibration of the Universe and tap the field of infinite possibilities. When you remove any fears you will begin to align with the Universe.

We can connect to the Universe using the principle of coherence. Coherence essentially implies the logical, orderly, and consistent relation of parts. As per physics, in a state of coherence, there is a total synchronization between the parts.

As we're a part of the Universe, the way to connect to the Universe is by increasing our coherence with the Universe. There are various ways we can use to align the frequency and vibrations of our body, mind, and heart with the frequency and vibrations of the Universe to create synchronicity. When you're "in sync", you can "entangle" with the Quantum Field.

We also have to believe in the power of belief. There's a link between mind and matter, and that all the cells of your body are affected by your thoughts. What you believe in affects your mind and body and they begin to perceive it as a reality. And if you're successful in synchronization with the Quantum Field, you manifest that reality.

If you believe that you're not connected with the Earth and the Quantum Field, then your belief system will not be able to vibrate at the frequency of the connection. So, you must believe that there is a connection and that you cannot be separate from the Earth and the Universe. Once you start believing, you come into alignment of vibration and frequency of the Quantum Field.

Here are the eight sequential steps you should take to connect yourself with the Quantum Field of infinite possibilities as per my personal experience.

(i) Sit Quietly

It's all about getting into the present. Sit feeling grounded with your feet on the ground. Pay attention to your breath as you breathe slowly in a natural rhythm.

(ii) Synchronize with the heart

As you feel into the body, hear your heartbeat. You've to come into coherence with your heart. The synchronization with the heart is the connection point with the Quantum Field.

(iii) Listen to the inner voice

Sitting comfortably, breathing in coherence with your heart, and with a quieter mind, you start to connect with the Quantum Field. You feel the breath and your heartbeat and as you tune-up with them, you start to listen to your inner voice coming from the depths of your heart.

(iv) Express your intentions

When you have the heart coherence and body grounding, you begin to see what you really want and what you're really asking for. Listen and see what comes forth in the image. Your intention could be food, shelter, resources, money, or even non-material needs like love and peace.

(v) Feel your intention

The feeling is critical – you have to "act as if". You need to have the acceptance that your intention has already been manifested and that you're already experiencing it. Shaman's call this "dreaming your world into being".

(vi) Synchronize with the Universe

After your heart alignment and grounding of the body with Earth, you come into synchronization with the Quantum Field. Your vibrations or energy ripples out of your body and you bring it back in so the energy of the Universe gets connected with you. You feel like your heart is lifted as you feel the joy and the buoyant positive energy.

(vii) Make it a routine

You should repeat the above-mentioned steps of this manifestation equation of creating the life you want daily for the next seven days first thing in the morning and the last thing in the night.

14.3   Why do we believe that we are all connected?

Although Krishna is situated in His own abode (goloka eva nivasaty akhilātma-bhūtaḥ (Brahma-Samhita 5.37)), where He enjoys His transcendental pastimes with the cowherd boys and gopīs, He is nevertheless present everywhere, even within the atoms of this universe (aṇḍāntara-stha-paramāṇu-cayāntara-stham (Brahma-Samhita 5.35)).

Every cell in our body, down to its constituent parts, is alive. They are all working in unison to complete their tasks. Not just our cells, but even the subatomic particles are constantly moving with energy that can be converted to mass. Is energy different from mass? Is shakti different from Shiva? Why do we call us living if we are made up of same atoms as non-living things, just in a sequence so as to make us more conscious than others?

I used to question why we could meditate on anything. Why is there no law of meditating on a name or an image, the way Swami Vivekananda would focus his meditation on a black spot Ramakrishna's image is what he focused on during his meditation, but over time, it started to fade.

"TAT TWAM ASI" When in **Chandogya Upanishad**, gives the answer. It translates to "That Thou Art" or "You Are That". The phrase represents the idea that the individual self (Atman) is ultimately identical to the ultimate reality or ultimate consciousness (Brahman). This concept is central to **Advaita Vedanta**, a philosophical and spiritual school of Hinduism, and expresses the idea of nondualism, or the idea that all things are ultimately one and the same.

In **Islam**, the idea of Tawhid, or the oneness of God, encompasses the belief that all things in the universe are interconnected and dependent on each other, and that everything ultimately leads back to the divine unity of God. The Quran also emphasizes the interconnectedness of all things, stating that everything in the universe has a purpose and that nothing is insignificant.

In **Christianity**, the concept of the interconnection of all things can be seen in the idea of the body of Christ, where each individual is considered a part of the greater whole, and in the teachings of Saint Francis of Assisi, who believed that all creatures, from animals to plants to inanimate objects, were interconnected and deserving of respect and care.

The **Hindu belief** that we are all connected stems from the idea that everything in the universe is made up of the same particles/energy. This means that every object is connected by a single vibratory field, also known as the Akasha(even Tesla referred to the philosophical etheric field as akasha), Logos (In Bible, Logos was used initially instead of "Word"), OM (Hinduism), Higgs field, dark energy, and so on in different religions and scientific concepts. It is the same sphere of energy that saints, Buddhas, yogis, mystics, priests, shamans, and seers have seen when they gaze within. This field is seen as the common thread that connects all religions and sciences, as everything in the universe is constantly changing and exchanging information with this field. You never see anything in its totality because it is made up of layer upon layer of vibration and it is constantly changing, exchanging information with Akasha. Through the constraints of thinking, observation is an act of creation. We create the appearance of solidity, of things, by identifying and naming them. "Once you label me, you negate me" stated the philosopher Sren Kierkegaard. You eliminate all the other things I may be by assigning me a name, a label. By pinning it down and identifying it, you lock the particle into becoming a thing while also generating it and defining its existence. In Hinduism, the whole cosmos is seen as dancing to the drum of Shiva Nataraja, the "lord of the dance." The idea is that everything is constantly vibrating and exchanging information with the Akasha, from subatomic particles to galaxies, stars, planets, and all life. This idea is supported by the strange phenomenon of quantum entanglement, where two subatomic particles can be intimately linked to each other even if separated by vast distances. What are we? A set of subatomic particles? In addition, the work of Dr. Masaru Emoto on water consciousness has shown that even water can be affected by human thoughts, sounds, and intentions. This has led to the question of how much our thoughts and intentions can influence the world around us. This belief is supported by both spiritual experiences and scientific discoveries, and helps to explain how everything in the universe is interconnected.

Slowly we are trying to understand things in quantum world which are stated in Vedas, Upanishads and Gita. Heisenberg used to say.. Quantum theory will not look ridiculous to people who have read Vedanta.

The Hindu belief that we are all connected is based on the idea that everything is manifestation of the divine consciousness and therefore everything in the universe is made up of the same particles/energy and is connected by a single vibratory field, the Akasha. This belief is supported by both spiritual experiences and scientific discoveries, and helps to explain how everything in the universe is interconnected.

According to Bhagavad-Gita 6.30:

yo māṁ pashyati sarvatra sarvaṁ cha mayi pashyati
tasyāhaṁ na praṇashyāmi sa cha me na praṇashyati

Meaning: For those who see Me everywhere and see all things in Me, I am never lost, nor are they ever lost to Me.

CHAPTER 15

DISPASSIONATE THINKING

According to Cambridge dictionary, 'dispassionate' is defined as:

"able to think clearly and make good decisions by not being influenced by emotions".

According to Merriam Webster dictionary, 'dispassionate' is defined as:

"not influenced by strong feeling especially, not affected by personal or emotional involvement".

15.1   Karma Yoga: Dispassionate Action

Karma Yoga is an action-based technique that helps set one on the path towards Moksha or salvation. The word Karma is derived from *kṛ*, meaning "doing or acting" and includes everything a person can perform (e.g., physical actions, speech, thinking). Yoga is derived from *yuj* meaning "to link". Thus Karma Yoga is the way for a person to link to God via actions. Note these actions are those that are performed consciously and with a free will. So actions performed subconsciously are not included in the definition of action in Karma Yoga.

The classic explanation of Karma Yoga is in the Bhagavad Gita when Arjuna has doubts when he is called to action on the battlefield. The Bhagavad Gita is a dialogue between Krishna and Arjuna, where Krishna explains how do act in accordance with Dharma. The most famous statement in the Bhagavad Gita (2.47) is:

karmaṇy-evādhikāras te mā phaleṣhu kadāchana
mā karma-phala-hetur bhūr mā te saṅgo 'stvakarmaṇi (2.47)

Meaning: You have a right to perform your prescribed duties, but you are not entitled to the fruits of your actions. Never consider yourself to be the cause of the results of your activities, nor be attached to inaction.

At the same time, doing nothing (inaction) is not acceptable.

That is, using the potential results of one's actions cannot be the motivation to act but at the same time because one has no right to the benefits does not mean one does nothing. Arjuna is explicitly told that anyone who uses resources without working is committing a sin. Hence one is obliged to be selfless and fulfil one's duty towards the world. Karma Yoga is nothing but techniques on how to perform actions solely as duty and not expecting any rewards.

Karma Yogis are motivated by altruism or empathy. They are not motivated by their self-interests, but motivated to increase the welfare of others. Note that, in many cases, this could be detrimental to themselves. The absence of desire for rewards of one's actions clearly suggests altruistic motives. The Bhagavad Gita states that one must work incessantly and doing nothing is not acceptable and is not ethical. But the actions must not produce self-desire in one's mind. If one's desires are not the focus, one will not expect any rewards. This is called Nishkama Karma, Karma (action) without Kama (desires). That is, one's actions are performed in a dispassionate fashion.

Here is another way of looking at Nishkama Karma. Research shows that hard work, more than achievement is correlated to happiness. Carrots are important but do not work for everyone. So hard work, leading to achievement is the best approach. This, again, is not necessarily Karma Yoga, which requires one not to focus on achievement. One needs to take many steps before Nishkama Karma can be attained. Initially, all of us are focused on getting personal benefits. One then gives up personal benefits by actively thinking that one should not seek them. Only when not expecting rewards is automatic, has one reached the stage of Nishkama Karma.

Note that Karma Yoga is not just about performing ethical actions. Even if all of one's actions are ethical, one is not necessarily on a path of Karma Yoga because one may be attached to the results. One may feel entitled to the results because one is ethical. For example, one might help a colleague at work to get promoted but expects gratitude. This is Karma (with a positive impact) but not the behavior of a Karma Yogi. It is important to detach oneself from the benefits to be on the path of Karma Yoga.

Moksha, or realizing Brahman, occurs when one realizes that one's true self is identical to everyone and everything else's true self. That is, one has realized the interconnectedness of all beings in the universe with oneself. In this context, a Karma Yogi's work will be mainly about one's obligation towards others. They perform these actions because it is the right course that will benefit society. There will be total absence of desire or rewards for oneself. Karma Yogis does not expect even appreciation, gratitude or admiration from the people who benefit from the actions. There is total lack of selfish desire and work itself is worship. In other words, to become a Karma Yogi, one must be on a path of responsibility towards others. The aim is to clear the mind of jealousy, greed and focus only on one's duty.

Practically, Karma Yoga wants one to focus on the responsibilities associated with a task. This is also *Yogah Karmasu Kausalam,* meaning that one has to seek perfection in one's work. Just because one is not interested in rewards does not mean one does a shoddy job. One must be keen, work intelligently and aspire to complete all tasks to the highest level of quality.

To be really **dispassionate**, one should have a balanced mind which can find happiness in the welfare of others. It also requires control over one's senses. This is not via pretense of renunciation or denial of sensory inputs. Rather it is a conscious attempt to prevent the mind from thinking of self-indulgence. One must remain undisturbed in all situations as one is devoted to one's work. One goes about doing duties without looking at other people's reactions (be they criticism or praise). One is self-absorbed in one's work. This will enable one to maintain a calm mind and to intensely work even when one is surrounded by turmoil.

The only way to avoid thinking of the fruits of one's action is to get rid of one's ego. Ego leads to attachment that binds us to desires. Karma Yoga says that one's ego and the resulting selfish actions separates one from one's true Self. By performing actions in a selfless manner, one can break down this separation. Karma Yoga tells one to release oneself from the chains of attachment via Nishkama Karma.

In short, the core idea behind Karma-Yoga is to be duty-oriented and not result-oriented. That is, one must follow one's duty even though it may be personally uncomfortable. Therefore, one must execute all work as a liability that needs to be discharged. One is not owed the benefits of the actions. The aim is *Loksangraha* (welfare of the society) with Nishkama karma (unattached action). This does not require any belief in God. One simply has to exercise one's free will and perform actions that are according to Dharma but being detached at all the time. Of course, if one believes in God, Karma Yogis believe that the fruits of their action are all for God.

15.2 Dispassion in Buddhism

The conditioned states of mind have definite and direct causes. This is often referred to as the law of conditionality or the law of "if this occurs then that results":

"And what is the noble method that is rightly seen and understood by discernment? From the arising of this comes the arising of that. When this isn't, that isn't. From the cessation of this comes the cessation of that. "This is the noble method that is rightly seen and rightly ferreted out by discernment." (Anguttara Nikaya 10.92)

When a mind is aflame with passion suffering ensues.

Stillness is the present moment experience of a mind settled in equanimity. Stillness is a condition of mind developed through understanding the Four Noble Truths. Equanimity is a state of mind free of reaction, without movement caused by passion. All things in the phenomenal world have causes arising in the phenomenal world.

When the mind is settled in dispassionate mindfulness the agitation of clinging and desire abandoned.

It is the nature of a conditioned mind to grasp. When thoughts become agitated by passion, grasping arises. Grasping (and the ensuing clinging) is the cause of stress and suffering known as dukkha. Dukkha arises from all forms of grasping: desire, craving, greed, and aversion. All arise from ignorance or a lack of wisdom. Ultimately, all created things arise from desire or passion, and all created things are put aside through dispassion. Dispassion arising from wisdom is the condition of an awakened mind. When passion ceases, cessation of suffering arises.

Passion arises from attaching a thought to a feeling creating emotion. Passion is the action or movement, physical, mental, or combined, arising from emotion: An experience provides pleasure or fulfillment in some manner and instead of simply and dispassionately experiencing the feeling, we decide that we want more of the pleasure-providing experience. Passion also arises from an experience that is perceived as unpleasant or unfulfilling and we decide that we do not want to have the experience again.

Every discriminating judgment of an experience, whether the judgment is "positive" or "negative" creates more desire or more aversion, more passion. This repetitive and discursive response to passion is the cause of conditioned thinking.

Dispassion is not negative passion. Dispassion is the expression of wisdom arising from present-moment mindfulness of the true nature of reality. Dispassion is not a mind state that can be generated through passion. Dispassion naturally arises through a practice of putting aside that which gives rise to passion.

"Among whatever qualities there may be, fabricated or un-fabricated, the quality of dispassion – the subduing of intoxication, the elimination of thirst, the uprooting of attachment, the interruption of discursive thinking, the destruction of craving, (develops) dispassion, cessation, the realization of unbinding (Nirvana) – this is considered supreme. Those who have confidence in the quality of dispassion have confidence in what is supreme; and for those with confidence in the supreme, supreme is the result." (Ituvattaka Sutta v90)

Dispassion is not detachment. Dispassion is deep mindful engagement. Dispassion is not arrogant and irreverent aloofness. Dispassion is mindful and complete acceptance. Dispassion is not avoidance. Dispassion is mindfulness of whatever is occurring as events occur free of discriminating thoughts. Dispassion is not based on ignorance. Dispassion is a mindful expression of true wisdom.

Dispassion does not generate inaction. Dispassion arising from wisdom leads to skillful and effective action. Dispassion is not grim stoicism. Dispassion generates skillful and compassionate acts informed by wisdom. Dispassion does not give rise to isolation. Dispassion arises from the wisdom and understanding of the interconnectedness of all things.

Dispassion gives rise to true understanding of Anatta and a life without the effects of providing for an ego-centered self. (See link below for an explanation of Anicca, Dukkha and Anatta) Dispassionately through wisdom gained from the Eightfold Path we cease making choices that satisfy the cravings of the impermanent ego-self. Initially this will often give rise to more passion. When the ego-self is denied that which it has always received, constant sense-fulfillment, the mental reaction can be quite passionate. Our minds will often thrash about, complaining through distraction, doubt, fear, confusion and sometimes even physical discomfort.

The Buddha likened establishing a meditation and spiritual practice to taming a wild elephant. In order for a young elephant to be useful, it must be able to focus and be aware of its true nature. To tame a young elephant, a strong rope would be tied around the elephant's neck and to a strong post or tree. The elephant would immediately begin thrashing around, flapping its ears, stomping the ground, and making loud grunts and bellows, very unhappy to not be able to wander around aimlessly engaging in any distraction that arose, feeding its passions. The more resistant the young elephant became, the stronger the rope held. Eventually the elephant would put aside its desire for continual distraction and sensual fulfillment and it would settle down. At a certain point the elephant will let go of its need to be anything other than what it is.

As we establish our own meditation practice, our minds are often passionately thrashing about, resistant to settling down. Thoughts insist on wandering aimlessly with strong desire to continue distraction by following one thought with another, continually describing their own self-created reality. Initially our minds have a great aversion to give up constant craving for physical and mental fulfillment. As we continue and deepen our meditation practice, our thoughts settle down.

In this metaphor, our mind is the young elephant, the rope is mindfulness or awareness of our breath, and the strong post or tree is our breath. As we use mindful awareness of our breath to disengage from our thoughts and settle our awareness on the present moment, putting aside passion, we become liberated and free. By utilizing the simple method of breath-awareness meditation we are able to tame our own wild minds. Once settled in stillness wisdom arises and dispassion is present.

By practicing all of the factors of the Eightfold Path we begin to minimize the effects satisfying and protecting the passions of the ego-self and begin to experience our true and dispassionate nature.

Returning to our metaphor, once the elephant has learned to remain mindful of the post, the rope is loosened and the elephant is finally free. Once we learn mindful awareness of our breath and integrate all of the factors of the Eightfold Path, passion subsides and true freedom arises.

The practice of awakening through the understanding of the Four Noble Truths by mindfully integrating the Eightfold Path unfolds in the phenomenal world as peaceful mindfulness of life as life occurs. Dispassion arising from wisdom brings ever deeper awareness of the calm and spacious nature of mind. Dispassion gives rise to the innate ability to bring virtue, concentration and wisdom into every situation, into every experience, without the distracting thought that our experience be any different than what is occurring.

The Buddha's way of dispassion leads to complete integration of psychic faculties: it gives the penetration vision that sees directly into the nature of causality, and beyond it, to the uncaused and uncompounded. That having been attained, no external events, no happenings in the realm of relative reality can give rise to sorrow, resentment or desire. The mind is finally liberated, posed on the wave crest of the ocean of saṃsāra, never to be submerged beneath the seething waters. Knowing this body to be as foam and understanding its mirage-like nature, one will escape the tight grip of the King of Death, having destroyed the power of Māra. No longer is there friend or foe for him who is thus liberated.

Dispassion is a calm and abiding ego-less presence.

Dispassion is the peace and freedom arising from unconditioned mind.

15.3   More on Dispassion

The concept of dispassion has been an important part of many different ethical traditions. Almost all of these traditions also have some emphasis on related concepts like detachment, mindfulness, and tranquility, and on practices like meditation, prayer, or contemplation. These notions are probably most readily connected by many of us with Buddhism, thanks no doubt in part to the attractions of Zen to the American counter-culture of the 1950s and 60s, the popularity and charisma of the current Dalai Lama, and the rise of positive psychology, with its consistent interest in mindfulness and meditation. But they are also present in Hinduism, and to a lesser (or at least less obvious) extent in Daoism and Confucianism. Further to the west, they were an important part of Stoic and Christian thinking about virtue, especially in the Christian spirituality that developed in the eastern part of the late Roman empire and which continues in today's Eastern Orthodox Church. In spite of this diverse and venerable pedigree, the idea of dispassion as an ethical ideal has received almost no attention from philosophers in the west, outside of specialists in ancient or Asian philosophy.

15.4   About Dispassion in Katha Upanishad

mṛtyuproktāṃ naciketo 'tha labdhvā
vidyāmetāṃ yogavidhiṃ ca kṛtsnam.
brahmaprāpto virajo 'bhūdvimṛtyu-
ranyo 'pyevaṃ yo vidadhyātmameva. . 2-III-18.

Meaning: 2-III-18. Nachiketas then, having acquired this knowledge imparted by Death, as also the instructions on Yoga in entirety, attained Brahman having become **dispassionate** and deathless. So does become anyone else also who knows the inner Self thus.

Measurement Techniques of Spiritual Intelligence

CHAPTER 16

MEASUREMENT OF SPIRITUAL INTELLIGENCE BY ISIS

16.1 Integrated Spiritual Intelligence Scale (ISIS)

Yosi Amram developed a grounded theory of Spiritual Intelligence based on interviews with 71 people of different religious and spiritual orientations nominated by their associates as embodying their spiritual values in daily life, including a minimum of four interviews with people representing the following traditions: Buddhism, Christianity, Earth-Based (Shamanic and Pagan) paths, Hinduism (including Yogic traditions), Islam/Sufism, Judaism, Non-Duality, and Taoism. The majority were spiritual teachers recognized in their lineages (rishis, priests, rabbis, swamis, sheikhs, etc.). Employing a grounded theory method, open coding identified individual properties demonstrated by participants (e.g., gratitude, joy, abundance, and appreciation of beauty), followed by axial coding to identify themes (e.g., love of life, which combined all the above). Selective coding identified higher-level themes, such as grace, which combined the themes of love of life, the sacred, and trust. Interviews continued until saturation was achieved, and the levels of significance for convergence of unprompted responses to neutrally worded questions were extremely high (the minimum accepted was 80%).

The model used to develop a validated instrument for measuring the Spiritual Intelligence, was called the **Integrated Spiritual Intelligence Scale (ISIS).**

The analysis produced **seven major themes** with subthemes universal across traditions considered to be the capacities that distinguish SI from other intelligences:

16.1.1 Consciousness—Developed refined awareness and self-knowledge:

(i)   **Mindfulness**—knowing self and living consciously with clear intention and mindful, embodied awareness and presence;

(ii)   **Trans-rational knowing**—transcending rationality through synthesis of paradoxes and using various states/ modes of consciousness (e.g., meditation, prayer, silence, intuition, and dreams) to access knowledge;

(iii) **Practice**—using a variety of practices to develop and refine consciousness or spiritual qualities.

16.1.2 Grace—Living in alignment with the sacred, manifesting love for, and trust in life:

(i)   **Sacred**—living in alignment with the divine, a universal life force, nature, or one's true essential nature;

(ii)   **Love of life**—reverence and cherishing of life based on gratitude, beauty, vitality, and joy;

(iii) **Trust**—hopeful/optimistic outlook based on faith or trust.

16.1.3 Meaning—Experiencing significance in daily activities through a sense of purpose and a call for service, including in the face of pain and suffering.

16.1.4 Transcendence—Going beyond the separate egoic self into an interconnected wholeness:

(i)   **Relational I-Thou**—nurturing relationships and community with acceptance, respect, empathy, compassion, loving-kindness, generosity, and I-Thou orientation;

(ii) **Holism**—utilize a systems perspective seeing the wholeness, unity, and the inter- connections among diversity and differentiation.

16.1.5 Truth—Living in open acceptance, curiosity, and love for all creation (all that is):

(i) **Acceptance**—forgive, embrace, and love what is, including the "negative" and shadow;

(ii) **Openness**—open heart and mind, open curiosity, including open respect for the wisdom of multiple traditions.

16.1.6 Serenity—Peaceful surrender to Self (Truth, God, Absolute, true nature):

(i) **Peacefulness**—centered, equanimity, self-acceptance, self-compassion, and inner- wholeness.

(ii) **Ego-lessness**—letting go of persona to maintain humble receptivity, surrendering, and allowing what wants and needs to happen.

16.1.7 Inner-Directedness—Inner-freedom aligned in responsible wise action:

(i) **Freedom**—liberation from conditioning, attachments and fears, manifesting courage, creativity, and playfulness;

(ii) **Discernment**—wisdom to know truth using an inner-compass (conscience);

(iii) **Integrity**—being/acting authentically, responsibly, and with alignment to one's values.

This model, integrated spiritual intelligence scale **(ISIS)** with 83 self-report items scored on a 6-point Likert-type scale of frequency, and later a validated 45-item short form, available in both self-report and as a 360-assessment measure, similar to other competency models assessing motive- and trait-level qualities.

ISIS features 22 subscales assessing separate SI capabilities related to (i) Beauty, (ii) Discernment, (iii) Ego-lessness, (iv) Equanimity, (v) Freedom, (vi) Gratitude, (vii) Higher-self, (viii) Holism, (ix) Immanence, (x) Inner-wholeness, (xi) Intuition, (xii) Joy, (xiii) Mindfulness, (xiv) Openness, (xv) Practice, (xvi) Presence, (xvii) Purpose, (xviii) Relatedness, (xix) Sacredness, (xx) Service, (xxi) Synthesis, and (xxii) Trust. These 22 subscales cluster into 5 domains: (i) Consciousness, (ii) Grace, (iii) Meaning, (iv) Transcendence, and (v) Truth, which overlap with five of the seven major themes identified in the qualitative research. Sample items include:

(i) **Consciousness-synthesis:** I can hold as true and integrate seemingly conflicting or contradictory points of view;

(ii) **Grace**—Living in alignment with the sacred, manifesting love for, and trust in life;

(iii) **Meaning-Service:** In my daily life, I feel my work is in service to the larger whole;

(iv) **Transcendence-Sacredness:** I live in harmony with a force greater than myself—a universal life force, the divine, or nature—to act spontaneously and effortlessly;

(v) **Truth-Openness:** I hold resentment towards those who have wronged me (reverse scored).

The overall internal consistency of the ISIS was high (Cronbach's alpha = 0.97), as was internal consistency of the domain scales (Cronbach's alpha range 0.84–0.95, M = 0.89). The internal consistency of the capability subscales was moderate to high (Cronbach's alpha range 0.62–0.88, M = 0.75).

The ISIS showed correlation with satisfaction with life. Furthermore, studies by other researchers, some of whom translated the ISIS into other languages, applying it in different cultural contexts and found it valid and reliable in Turkish and

Persian. Additionally, other studies using the ISIS also found it not only to be valid and reliable, but also correlated with a variety of positive outcomes in different cultural and religious contexts, including:

(i)   Leadership effectiveness as rated by followers in the US;

(ii)   Resilience among Christians in India;

(iii)   Leadership effectiveness among Turkish managers as measured by financial performance of their organizations;

(iv)   Job satisfaction among teachers in Iran;

(v)   Work performance among nurses in Malaysia;

(vi)   Work satisfaction among employees in Slovenia;

(vii)   Organizational performance as measured by return on assets and another financial outcome measure among banks in Pakistan;

(viii) Mental health in India; and

(ix)   Life satisfaction and awe in China.

In some of the above-mentioned studies which controlled for other established constructs, SI provided incremental predictive validity. For example, it was found that high scores on ISIS predicted death anxiety among Iranian women when controlling for social intelligence. Further, it was found SI to predict organizational financial performance of leaders, while emotional intelligence did not. Furthermore, an author overcame the limitations of common-method bias and its halo-effect and found that the self-reported spiritual intelligence was associated with improved organization financial performance, measured independently and objectively. Similarly, it was found that self-reported SI among bank employees predicted the bank's financial performance as measured by their return on assets.

It was found that the ISIS self-report measure of spiritual intelligence predicted leadership effectiveness ratings by outside observers even after controlling for personality (using the Five Factor Model, 10-item personality inventory [TIPI]; and Emotional Intelligence. Self-reported EI and SI showed moderate correlation (RS = 0.27), which might be expected since both constructs feature similar competencies, such as self-awareness (EI) and self-knowledge (SI), and emotional self- regulation (EI) and equanimity (SI). Additionally, Amram deployed **360-assessments** of SI and EI, finding observer ratings of SI and EI each on their own to predict leadership effectiveness ratings from other observers. In analyzing the employee-assessment scores, EI alone explained 41% of leadership effectiveness while SI, on its own, explained 46% of leadership effectiveness. Combined, they explained 67% of leadership effectiveness. These results suggest that EI and SI complement each other and uniquely contribute to leadership effectiveness, with SI providing differential and incremental predictive validity.

In the ISIS-based measurement, research participants were 263 adult volunteers. The modal age category across the four samples was 35: 44, with 83 participants reporting this range; 1 participant reported an age of 17 or younger (and was excluded from analyses). The high spiritual intelligence and business acumen sample included 15 participants who completed the ISIS out of 18 who met the criteria and were invited. Of these 15 participants, 9 were men, 5 women, and 1 who did not report gender. The high spiritual intelligence and average business acumen teacher sample included 17 participants who completed the ISIS out of 27 who met the criteria and were invited to participate. Of these 17 participants, 5 were men, 7 women, and 5 did not report gender. All participants completed a battery of questionnaires, including the newly developed ISIS, the Satisfaction With Life Scale, the Index of Core Spiritual Experiences, and a brief demographic questionnaire assessing age and gender. All dependent variables were analyzed to determine whether any significant differences existed in responses to the two formats; no significant differences were found, and we excluded this factor from subsequent analyses.

16.2 ISIS Survey Questionnaire/Statements

ISIS is an 83-item long form, and a 45-item short form, self-report instrument.

Following is the ISIS Survey Questionnaire/Statements:

On the next few pages, please score all the items on a scale from 1 to 6 based on the general frequency of your behavior over the past 6 to 12 months:

(1) Never or almost never; (2) Very infrequently; (3) Somewhat infrequently; (4) Somewhat frequently; (5) Very frequently (6) Always or almost always.

1. I notice and appreciate the beauty that is uncovered in my work.

2. I expect the worst in life, and that's what I usually get.

3. When things are chaotic, I remain aware of what is happening without getting lost in my experience.

4. During an activity or conversation, I monitor and notice my thoughts and emotions.

5. I practice inner and outer quiet as a way of opening myself to receive creative insights.

6. I have a good sense for when my purpose requires nonconformity, out-of-the-box thinking, or taking an unpopular stand.

7. I resist events that I don't like, even when they need to occur.

8. In my daily life, I feel the source of life immanent and present within the physical world.

9. *I get upset when things don't go the way I want them to go.

10. In my day-to-day activities, I align my purpose with what wants to and needs to happen in the world.

11. I find it frustrating when I don't know what the truth is.

12. I pay attention to my dreams to gain insight to my life.

13. *In my daily life, I am disconnected from nature.

14. Seeing life's processes as cyclical rather than linear gives me useful insights to daily challenges.

15. A higher consciousness reveals my true path to me.

16. I live and act with awareness of my mortality.

17. *In difficult moments, I tap into and draw on a storehouse of stories, quotes, teachings, or other forms of time-proven wisdom.

18. *I don't know how to just be myself in interactions with others.

19. I hold my work as sacred.

20. *I have a daily spiritual practice such as meditation or prayer that I draw on to address life challenges.

21. I enjoy the small things in life such as taking a shower, brushing my teeth, or eating.

22. *I am driven and ruled by fears.

23. *I tend to think about the future or the past without attending to the present moment.

24. *My life is a gift, and I try to make the most of each moment.

25. *I draw on my compassion in my encounters with others.

26. I am limited in my life by the feeling that I have very few options available to me.

27. I spend time in nature to remind myself of the bigger picture.

28. *My actions are aligned with my values.

29. *In meetings or conversations, I pause several times to step back, observe, and re-assess the situation.

30. I use objects or places as reminders to align myself with what is sacred.

31. *I have a hard time going against conventions, expectations, or rules.

32. *Even when things are upsetting and chaotic around me, I remain centered and peaceful inside.

33. I find it upsetting to imagine that I will not achieve my desired outcomes.

34. *In my day-to-day tasks, I pay attention to that which cannot be put into words, such as indescribable sensual or spiritual experiences.

35. *I am aware of a wise- or higher-self in me that I listen to for guidance.

36. *I can hold as true and integrate seemingly conflicting or contradictory points of view.

37. *I strive for the integration or wholeness of all things

38. *My work is in alignment with my greater purpose.

39. *I derive meaning from the pain and suffering in my life.

40. I feel that my work is an expression of love.

41. I use rituals, rites, or ceremonies during times of transition.

42. My actions are aligned with my soul my essential, true nature.

43. I remember to consider what is unspoken, underground or hidden.

44. *Because I follow convention, I am not as successful as I could be.

45. I am aware of my inner truth what I know inside to be true.

46. *Being right is important to me.

47. *I notice and appreciate the sensuality and beauty of my daily life.

48. *I enhance my effectiveness through my connections and receptivity to others.

49. *Even in the midst of conflict, I look for and find connection and common ground.

50. *I listen to my gut feeling or intuition in making important choices.

51. *I listen deeply to both what is being said and what is not being said.

52. *I am mindful of my body's five senses during my daily tasks.

53. I seek to know what is logically provable and ignore the mysterious.

54. *I look for and try to discover my blind spots.

55. I have a hard time integrating various parts of my life.

56. I work toward expanding other peoples' awareness and perspectives.

57. *I live in harmony with a force greater than myself a universal life force, the divine, or nature to act spontaneously and effortlessly.

58. *My goals and purpose extend beyond the material world.

59. I draw on deep trust or faith when facing day-to-day challenges.

60. *I hold resentment towards those who have wronged me.

61. I feel like part of a larger cosmic organism or greater whole.

62. *I find ways to express my true self creatively.

63. When looking at others, I tend to focus on what they need to do to improve.

64. Experiences of ecstasy, grace, or awe give me insights or direction in dealing with daily problems.

65. *To gain insights in daily problems, I take a wide view or holistic perspective.

66. *I have daily and weekly times set aside for self-reflection and rejuvenation.

67. *I remember to feel grateful for the abundance of positive things in my life.

68. *I have faith and confidence that things will work out for the best.

69. I accept myself as I am with all my problems and limitations.

70. To solve problems, I draw on my ability to hold, accept and go beyond paradoxes.

71. *In my daily life, I feel my work is in service to the larger whole.

72. In arguing or negotiating, I am able to see things from the other person's perspective, even when I disagree.

73. *I see advancing my career as the main reason to do a good job.

74. I see financial rewards as being the primary goal of my work.

75. *My mind wanders away from what I am doing.

76. I am frustrated by my inability to find meaning in my daily life.

77. *Even when I seem to have very few choices, I feel free.

78. *I want to be treated as special.

79. *I have a hard time standing firm in my inner truth what I know inside to be true.

80. *I bring a feeling of joy to my activities.

81. *I strongly resist experiences that I find unpleasant.

82. *I am my own worst enemy.

83. I have answered all the questions truthfully and to the best of my ability

*** Items marked with * indicate items included in the short-form version (45-questions) of the ISIS.**

In the tests of the ISIS, evidence was found for its reliability, internal consistency, and temporal stability. Evidence was found for the convergent validity of the ISIS. In this theoretical framework, spiritual intelligence draws on spirituality, and spiritual intelligence predicts wellbeing. Therefore, the predicted positive correlations between the ISIS scores and the other scores. Moreover, consistent with this theoretical framework and despite other research to the contrary, these results suggest no direct relationship between spirituality and wellbeing when controlling for spiritual intelligence. Altogether, the results supported the convergent validity of the ISIS.

CHAPTER 17

MEASUREMENT OF SPIRITUAL INTELLIGENCE BY SISRI-24

17.1 Spiritual Intelligence Self-Report Inventory (SISRI-24)

D.B. King, in a master's thesis, defined spiritual intelligence as "a set of mental capacities which contribute to the awareness, integration, and adaptive application of the nonmaterial and transcendent aspects of one's existence, leading to such outcomes as deep existential reflection, enhancement of meaning, recognition of a transcendent self, and mastery of spiritual states". Subsequent research proposed a four-factor model for spiritual intelligence with capacities for: critical existential thinking, personal meaning production, transcendental awareness, and conscious state expansion. From this model King and Decicco (2009) developed and validated the spiritual intelligence self-report inventory (SISRI-24) with high internal reliability, test-retest reliability, and scale validity.

It was found that the SISRI-24 more closely related to intrinsic than extrinsic religiosity, and that the SISRI-24 Personal Meaning Making subscale was associated with greater personal meaning, mystical experiences, while the SISRI-24 Conscious State Expansion subscale was correlated with mystical experiences. King found a moderate correlation of RS=.4 between SI and EI when using the SISRI-24.

The SISRI-24 is arguably the most used of the SI inventories, perhaps for the convenience of its short length, and it has been translated into other languages and used cross-culturally. Studies by other researchers applying the SISRI-24 found it to be valid, reliable, and associated with, among other positive outcomes:

(i) Increased resilience and lowered perceived stress in Iran;

(ii) Greater mindfulness and transformational leadership among higher public education leaders in the US;

(iii) Higher EI, psychological ownership, caring behaviors, and reduced burnout among nurses in Malaysia;

(iv) Improved work performance among software company employees in India;

(v) General health and happiness among university students in Iran;

(vi) Student adjustment in Malaysia;

(vii) Enhanced meta-personal self-construal in Hong Kong; and

(viii) Greater self-care among hemodialysis patients in Iran.

However, recently some researchers attempted to determine whether a "general factor of spiritual intelligence" could be identified as a psychological construct by administering the SISRI to a large (N = 833) Polish adults. Their assumptions were that the SISRI-24 model has a hierarchical structuring, with 4 first-order factors (critical existential thinking, personal meaning production, transcendental awareness, and conscious state expansion, the original four factors of King's model loading onto a single second-order factor. Their results showed no support for a single factor of SI, and that previous studies —including the original one by King et. al.—were "highly problematic" in not showing strong support for the four-factor model underlying the SISRI-24. Specifically, some researchers found that development of the original scale (King's) did not follow the assumptions for King's four-factor model, and that initial analysis produced six factors. A number of items measured more than one factor and were later removed, though King claimed that the four-factor structure was "very well-supported" in the original 84-item pool. Some researchers further criticized the process of creating the original item pool and its shaping, showing that "the measurement is not fully derived from the theory but is molded ad hoc based on data" and that the theory never showed that the four-factor model measures a single higher-order factor of SI. They asserted that subsequent adaptations of the SISRI reflect these developmental problems, and that their testing of King's hypothesis gave an inadmissible solution, leading some researchers to conclude that "to date, no data support a single factor of SI measured by SISRI-24. . . and that measurement with this scale is highly problematic".

Whether a single, universal factor exists may or may not be valid or useful in considering SI as an "intelligence," these problems with the SISRI-24 model are worth further investigation, especially relative to the ISIS, which was developed in a very different way, as well as some other instruments that have been used in other cultural contexts.

17.2 Spiritual Intelligence Self-Report Inventory (SISRI–24)

The following statements are designed to measure various behaviors, thought processes, and mental characteristics. Read each statement carefully and choose which one of the five possible responses best reflects you by circling the corresponding number. If you are not sure, or if a statement does not seem to apply to you, choose the answer that seems the best. Please answer honestly and make responses based on how you actually are rather than how you would like to be.

The five possible responses are:

0 – Not at all true of me |

1 – Not very true of me |

2 – Somewhat true of me |

3 – Very true of me |

4 – Completely true of me

For each item, circle the one response that most accurately describes you.

1. I have often questioned or pondered the nature of reality. 0 1 2 3 4

2. I recognize aspects of myself that are deeper than my physical body. 0 1 2 3 4.

3. I have spent time contemplating the purpose or reason for my existence. 0 1 2 3 4

4. I am able to enter higher states of consciousness or awareness. 0 1 2 3 4

5. I am able to deeply contemplate what happens after death. 0 1 2 3 4

6. It is difficult for me to sense anything other than the physical and material. 0 1 2 3 4

7. My ability to find meaning and purpose in life helps me adapt to stressful situations. 0 1 2 3 4

8. I can control when I enter higher states of consciousness or awareness. 0 1 2 3 4

9. I have developed my own theories about such things as life, death, reality, and existence. 0 1 2 3 4

10. I am aware of a deeper connection between myself and other people. 0 1 2 3 4

11. I am able to define a purpose or reason for my life. 0 1 2 3 4

12. I am able to move freely between levels of consciousness or awareness (from waking consciousness which is comprised of our perceptions, memories, and thoughts to the superconscious state 0 1 2 3 4

13. I frequently contemplate the meaning of events in my life. 0 1 2 3 4

14. I define myself by my deeper, nonphysical self. 0 1 2 3 4

15. When I experience a failure, I am still able to find meaning in it. 0 1 2 3 4

16. I often see issues and choices more clearly while in higher states of consciousness/awareness. 0 1 2 3 4

17. I have often contemplated the relationship between human beings and the rest of the universe. 0 1 2 3 4

18. I am highly aware of the nonmaterial aspects of life. 0 1 2 3 4

19. I am able to make decisions according to my purpose in life. 0 1 2 3 4

20. I recognize qualities in people which are more meaningful than their body, personality, or emotions. 0 1 2 3 4

21. I have deeply contemplated whether or not there is some greater power or force (e.g., god, goddess, divine being, higher energy, etc.). 0 1 2 3 4

22. Recognizing the nonmaterial aspects of life helps me feel centered. 0 1 2 3 4

23. I am able to find meaning and purpose in my everyday experiences. 0 1 2 3 4

24. I have developed my own techniques for entering higher states of consciousness or awareness. 0 1 2 3 4

25. I have answered all the questions truthfully and to the best of my ability. 0 1 2 3 4

17.3   SISRI–24 Scoring Procedures

Total Spiritual Intelligence Score: Sum all item responses or subscale scores (after accounting for *reverse-coded item). 24 items in total; Range: 0–96.

Factors/Subscales:

I. Critical Existential Thinking (CET): Sum items 1, 3, 5, 9, 13, 17, and 21.

7 items in total; range: 0–28

II. Personal Meaning Production (PMP): Sum items 7, 11, 15, 19, and 23.

5 items in total; range: 0–20

III. Transcendental Awareness (TA): Sum items 2, 6*, 10, 14, 18, 20, and 22. 7 items in total; range: 0–28

IV. Conscious State Expansion (CSE): 116 Sum items 4, 8, 12, 16, and 24. 5 items in total; range: 0–20

*Reverse Coding: Item # 6 (response must be reversed prior to summing scores).

Higher scores represent higher levels of spiritual intelligence and/or each capacity.

17.4   Other Considerations of Measurement of SI

Whichever measures are used in different applications, SI has largely been shown to be predictive of positive outcomes, and evidence suggests that it is not equivalent to EI, though the two constructs overlap in some areas, such as self-awareness and self-knowledge. It is worth noting that even within an existential framework, some puzzling findings have arisen. For example, findings with ISIS showed a positive correlation with death anxiety; it was found higher scores on the SISRI-24 associated with increased death anxiety as well as dissociation with post-traumatic stress disorder in Greek first responders, and some researchers found a positive correlation between SI and death anxiety among Iranian veterans. Based on prior research on the association between spirituality and lower death anxiety, one might have expected that higher SI scores should correlate with lower death anxiety, and that is indeed what was found by others: high scores on the SISRI-24 were associated with lower death anxiety and higher spiritual wellbeing among Iranian veterans.

Furthermore, two Pakistani authors found a negative correlation between life satisfaction and the "meaning" aspect of SI among elderly Pakistani Muslims, leading the authors to suggest that possessing meaning and searching for meaning may make people momentarily aware of living a purposeless life, though many continue to believe their lives are purposeful and meaningful. In the same study, life satisfaction was correlated with consciousness, grace, meaning, truth, and transcendence. Nevertheless, the authors asserted that "spiritual intelligence can be applied to every aspect of daily life to experience greater meaning and well-being by exercising abilities such as mindfulness, presence, and equanimity even when an individual is experiencing pain and suffering".

Many studies have examined SI outside of an explicitly spiritual or existential context, and they also have shown mixed results. For example, some researchers found significant positive correlation between scores on the SISRI-24 and employees' organizational citizenship behaviors (willingness to contribute discretionary effort) but no significant relationship to the quality of their work performance in Islamic banks in Indonesia. Similarly, a researcher found positive correlations between high SISRI-24 scores and self-efficacy and passion for inventing and founding a new venture among Iranian entrepreneurial owner-managers, but not with the passion for developing the sustainability of a business. Another study) found SI to have an insignificant effect on leadership competence, especially when compared to EI and social competency in a sample of 900 respondents of an Airport Authority.

Interestingly, a study (in 2022) of Aotearoa New Zealand university students showed that SI was significantly associated with increased resilience and reduced depression and stress, but not anxiety; however, resilience scores fully mediated the combined depression, anxiety and stress scores. Therefore, it may only be that in some domains that SI "acts as a moderator in many situations but is not directly influential".

Clearly more research is needed to examine these relationships, especially in light of questions about the SISRI-24 model (King and DeCicco 2009), and what also may be language and cultural differences. For instance, the SISRI items contain words like "consciousness" and "meaning" in English, which may be difficult to render comparably in different languages and cultures, and some cultures value social desirability and conformity more than others, potentially confounding attempts at cross-cultural, cross-lingual comparisons. Even the titles of SI inventories may produce social desirability bias, always a problem in any self-report inventory. Further, since a large number of the published correlational studies involve Islamic populations and relatively fewer studies have assessed the validity of SI models in Buddhist, Taoist, Latinx and Indigenous cultures, much remains to be done to assess the universality of any of these models of SI, though it is possible that the Amram and Dryer (2008) model may provide the most rational starting place.

CHAPTER 18

OTHER METHODS OF MEASURING THE SPIRITUAL INTELLIGENCE

18.1   A 29-item Spiritual Intelligence Questionnaire (Abdollahzadeh)

A 29-item Spiritual Intelligence Questionnaire was developed and normed against university student populations in Iran.

This test was normalized by Abdollahzadeh in 2008 with the collaboration of Mahdieh Kashmiri and Fatemeh Arabameri on students. The normal group was 280 people, 200 of whom were the students of Gorgan University of Natural Resources and 80 students of Payame Noor University of Behshahr. Of these, 184 were female and 96 were male. First, a 30-item questionnaire was prepared by the test developers and implemented on 30 students. The Test reliability initially was 0.87 Cronbach's alpha, but the revised version garnered a score of 0.89. Factor analysis produced two major factors, "understanding and communicating with the source of the universe" and "spiritual life or reliance on the inner core". Both factors were correlated with Amram and Dryer's ISIS and demonstrated convergent validity. Using this instrument, Mohammadyari found that Iranian children of parents with greater levels of spiritual intelligence had higher levels of mental health, and Bolghan-Abadi et al. found this measure of SI to predict quality of life among university students in Iran.

18.2   The 29-item Questionnaire

Following is the 29-item questionnaire developed by Abdollahzadeh:

(with the responses against each item as (i) completely agree, (ii) agree, (iii) almost agree, (iv) disagree, and (v) completely disagree.)

1. I become astonished by observing the universe.

2. I am interested in searching and asking basic questions about life and universe.

3. I want to have a humane and compassionate relationship with others.

4. I always feel that God is watching over my actions.

5. I have a sense of gratitude and thanksgiving in life.

6. I live with enthusiasm.

7. I believe in God's divine presence in the world.

8. It is pleasing to me to pray and I feel calm after the worship.

9. In the face of difficulties and suffering, I believe that God helps me.

10. I can express my mistakes with regard to my position.

11. I pray and make efforts while facing problems.

12. I feel responsible and committed to my duties.

13. I have the ability to stand up against public in the event of opposition to the fundamental principles of life.

14. I enjoy helping others.

15. I do not forget God if I feel desperate.

16. I feel God's love to myself, both directly and through others.

17. I am not vulnerable to changes in the world because I believe the world is changing and has the ability of flexibly.

18.   I consider myself as the cause of all my feelings.

19.   My life is meaningful with a sense of value and purpose.

20.   I establish a spiritual connection with the person I help.

21.   I control my thoughts and actions and try to improve my development.

22.   I enjoy religious and spiritual foundations as a source of power and guidance.

23.   I believe that there is peace, love, joy and satisfaction within me, not in the world around me.

24.   I feel I am connected to the source of the universe.

25.   I feel secure on my own inner strengths and characteristics.

26.   I consider work as a tool for creativity and self-confidence (not just for money).

27.   I believe that I have nothing to lose because God is the real owner of everything.

28.   I have the ability to love and forgive others, regardless of gender, race or nationality.

29.   I find happiness and perfection in the light of attention to perfections and spirituality.

Two spiritual intelligence scales have been developed in India with reference to Hindu philosophy and values.

18.3   Dhar and Dhar's 53-item Spiritual Intelligence Scale

Dhar and Dhar (2010) developed an SI instrument of a 53-item Spiritual Intelligence Scale standardized on business executives on business executives that features **six subscales**: (i) benevolence, (ii) modesty, (iii) conviction, (iv) compassion, (v) magnanimity, and (vi) optimism. Split-half reliability corrected for full length using the Spearman-Brown formula on data from the test sample of 323 participants yielded a reliability coefficient of 0.98. It has been used in several educational and workplace settings in India on youth and adults, with findings that support the contribution of SI to various positive outcomes. Kumar and Mehta developed a 20-item spiritual intelligence scale consisting of **six factors:** (i) purpose in life, (ii) human values, (iii) compassion, (iv) commitment towards humanity, (v) understanding self and (vi) conscience with a Cronbach alpha and split-half reliability of 0.78. They used it and a measure of EI to assess 450 tenth-grade adolescent boys for educational adjustment and found both SI and EI to predict educational adjustment. Some other researchers used this model and Scale for Spiritual Intelligence (SSI), originally developed from Indian philosophy, adapting it into Turkish and confirmed it to have a six-factor structure showing internal consistency and reliability and was associated with greater meaning in life.

It has been used in several educational and workplace settings in India on youth and adults, with findings that support the contribution of SI to various positive outcomes.

18.4   A 20-item Spiritual Intelligence Scale

Kumar and Mehta (2011) developed a 20-item spiritual intelligence scale consisting of six factors: purpose in life, human values, compassion, commitment towards humanity, understanding self and conscience with a Cronbach alpha and split-half reliability of 0.78. They used it and a measure of EI to assess 450 tenth-grade adolescent boys for educational adjustment and found both SI and EI to predict educational adjustment. Erduran-Tekin and Halil (2019) used the Kumar and Mehta Scale for Spiritual Intelligence (SSI), originally developed from Indian philosophy, adapting it into Turkish and confirmed it to have a six-factor structure showing internal consistency and reliability and was associated with greater meaning in life.

18.5 Islamic Spiritual Intelligence Measurement (ISIM)

Significant research on SI has been conducted in Muslim contexts and perhaps using scales developed especially for those cultures. For example, some studies have spiritual intelligence scales developed in Muslim cultures about whose development little published information is available, but which have been reported to show correlations with positive, successful outcomes in problem-solving professional contexts. Recently, an Islamic Spiritual Intelligence Measure **(ISIM)** was recently derived from factor analysis using a sample of Muslim postgraduate students in Malaysia with **eight factors**: (i) God consciousness, (ii) trust in God, (iii) repentance, (iv) patience, (v) truthfulness, (vi) fairness, (vii) integrity, and (viii) continuous learning. It was synthesized from Islamic and Western perspectives, consisting of **seven themes**: (i) meaning/purpose in life, (ii) consciousness, (iii) transcendence, (iv) spiritual resources, (v) self-determination, (vi) reflection-soul purification, and (vii) spiritual coping with obstacles.

18.6 A 34-item Spiritual Intelligence Scale

Feng et al. argued that the spiritual intelligence of those influenced by Confucianism, Taoism, and Buddhism in China (specifically) should differ from Western models of SI. They interviewed and surveyed 50 people engaged in secular occupations, some of whom were nominated as "individuals who did meaningful things" to collect behavioral event statements, which were sorted by expert raters, to produce a Spiritual Intelligence Scale comprising 34 items rated on a 5-point Likert-type scale administered through an on-line survey to students and employees in mainland China. Factor analysis produced a three-factor model comprising Identification of Meaning (individuals determine the positive effects of their own experience and determine positive effects through perceptions of others and the outside world), Connection of Meaning (commonalities between different communities), and Realization of Meaning (realization of society, which apparently means self-sacrifice for the greater good, and realization of nature. The secular emphasis of the authors' approach throughout the conduct of their research runs counter to the burgeoning of Chinese folk religions since the relaxation of state sanctions regulating religion, including the revival of ancestor worship, shamanism, etc. and estimates that over 85% of the Chinese people believe in the supernatural. It seems likely that Feng et al., were attempting to create a secular version of SI congruent with the government's stance but not necessarily representative of the transcendent SI qualities many Chinese people, especially spiritual exemplars, may exhibit.

18.7   Spirituality Index of Well-Being

Daaleman, T. P. & Frey, B. B. (2004) have proposed a spirituality index, called, "Spirituality Index of Well-Being (SWIB).

It is a 12-item scale, with the scores given as under:

Strongly agree: 1, Agree: 2, Neither agree, nor disagree: 3, Disagree: 4, Strongly disagree: 05.

Given below are these 12 items:

1. There is not much I can do to help myself.

2. Often, there is no way I can complete what I have started.

3. I can't begin to understand my problems.

4. I am overwhelmed when I have personal difficulties and problems.

5. I don't know how to begin to solve my problems.

6. There is not much I can do to make a difference in my life.

7. I haven't found my life's purpose yet.

8. I don't know who I am, where I came from, or where I am going.

9. I have a lack of purpose in my life.

10. In this world, I don't know where I fit in.

11. I am far from understanding the meaning of life.

12. There is a great void in my life at this time.

18.8 A 13-item Islamic Spiritual Intelligence (ISI) Scale

1. I often contemplate about the relationship between men and Allah.

2. I believe that work is a religious obligation.

3. I attend sermons and prayers to enhance my knowledge about Islam.

4. I am able to emphasize Islamic values and ethics in my daily routine.

5. I believe that fulfilling Islamic obligation as a priority than making money.

6. I make time for my subordinates to consult with me at any time.

7. I encourage employees to voice out their opinions to me.

8. I have no problems to tell my subordinates what is right and wrong according to Islam.

9. I am totally forthright delivering message to my subordinates to adhere with Islamic values.

10. I can anticipate problems before these arise.

11. I turn to Allah when I cannot solve any problems.

12.   I always try to find new ways or methods to run the organization better.

13.   I consider myself as someone who is dynamic in thinking and making decision

18.9   Scale of Spiritual Intelligence (SSI)

Canales and Huaman (2020) has developed an 18-item Scale of Spiritual Intelligence (SSI), with the following weightage for each item:

Nothing True for me (score: 0), Somewhat True for me (score: 1), Quite True for me (score: 2), Totally true for me (score: 3).

Following is the list these 18 items:

1.   I think that in life everything has a deep meaning.

2.   No matter the place or circumstance, I always act according to my principles.

3.   My moments of spiritual practice renew my physical strength.

4.   When I think of the miracle of my existence, I am filled with joy.

5.   When I am dedicated to the noble mission of my work practice (profession) my forces multiply.

6.   In my free time I enjoy nature.

7.   My mind calms down when I reflect on some spiritual text.

8.   I believe that caring for my body and underprivileged is a sacred duty.

9.   When I experience failure, I can still find meaning in it.

10.   I often reflect on the meaning of events in my life.

11.   When someone needs me, I always take time to help.

12. I define myself by my deeper being and not by my physical being.

13. I am able to reflect deeply on what may be beyond death.

14. Beyond the human plane, there is a higher Being with whom we can relate.

15. I often see situations and options more clearly when I meditate, or pray.

16. I am aware that there is a deeper connection between myself and other people.

17. I am sure that helping others is my mission in this life.

18. There is definitely something beyond the physical and material world.

18.10   Spiritual Intelligence Scale for Children

Grasmane, et.al. (2022) produced a children spiritual intelligence scale with the following constructs: creation and awareness of personal meaning, self-understanding, mastery of self-control, awareness of personal authenticity and uniqueness, and social mastery. Their model was based on earlier models of SI, research on children's spirituality, EI, and integral theory. Expert review and piloting with young children improved the face validity of items, which were later tested on primarily male children in grades 1–4 in Lativia. Factor analysis and tests for internal consistency produced the final 23-item scale with acceptable psychometric qualities.

SECTION-IV

Application and Benefits of Spiritual Intelligence

CHAPTER 19

APPLICATION AND BENEFITS OF SPIRITUAL INTELLIGENCE

19.1   Application of Spiritual Intelligence

Spiritual intelligence is needed to find and to use the innermost resources, such as the capacity to care for, and the strength to tolerate and to adapt; developing a clear and stable sense of identity as an individual in the context of shifting work relations; the ability to discern the true meaning of the situation and to create meaningful work; identifying and harmonizing personal values with a clear goal; applying values without compromise to show integrity and the ability to restrain the ego before restraining creativity and helping employees to make their work more meaningful, and helping them reduce their ego through self-reflection. In turn, this will increase the individual potentials and the organization's work performance.

Whichever measures are used in different applications, the spiritual intelligence has largely been shown to be predictive of positive outcomes, and evidence suggests it is not equivalent to EI (Emotional Intelligence), though the two constructs overlap in some areas, such as self-awareness and self-knowledge. Spiritual intelligence can be applied to every aspect of daily life to experience greater meaning and well-being by exercising abilities such as mindfulness, presence, and equanimity even when an individual is experiencing pain and suffering. Many studies have examined SI outside of an explicitly spiritual or existential context, and they also have shown useful results.

A study (2022) with the university students showed that SI was significantly associated with increased resilience and reduced depression and stress, but not anxiety; however, resilience scores fully mediated the combined depression, anxiety and stress scores. Therefore, it may only be that in some domains SI "acts as a moderator in many situations but is not directly influential.

An approach (1995, 1999) to the intelligences, like identification of motive- and trait-level competencies that predict superior performance, seems to refer mainly to the strength of innate traits differentially distributed across the population. However, even traits can be developed, just as native cognitive intelligence (IQ), for example, can be honed by training in logic, strategy, analogies, and other forms of reasoning and problem-solving. So, too, can Spiritual Intelligence, at least as assessed by some of the techniques. Specific training in SI constructs produces measurable differences.

In healthcare, some researchers, have tried to develop SI as way to help people cope with existential challenges. For example, two independent studies (2020) showed that training in Spiritual Intelligence and subsequent assessment using Amram and Dryer's ISIS and King's SISRI-24, respectively, have a positive effect on hope and life expectancy among the chronically ill. Others have trained care providers in SI to work with patients. It was found that SI training improved nurses' competence in spiritual care in critical care units, which in turn improved patient quality of life. And SI training interventions have been shown to improve the quality of work life among healthcare providers. A metanalysis (2022) of seven studies comprising 512 nurses and nursing students showed that those who received SI training demonstrated higher SI scores at two- and four-week follow-up compared to controls, and that the experimental group scored significantly higher in communication skills, job satisfaction, and spiritual care competence and significantly lower in overall stress compared to controls. Significantly higher job satisfaction was reported at two-month follow-up by those who had had SI training.

Research conducted outside a healthcare setting examined whether mental health generally could be improved through SI training. In a study (2014) of the effect of SI training on the mental health of high school students, the intervention decreased interpersonal sensitivity, somatization, obsessive-compulsive behavior, depression, anxiety, aggression, phobia, paranoid ideation, and psychoticism in the experimental group compared with controls.

Thus, it appears that like other intelligences, innate SI capacities, which theoretically should be differentially distributed in a population like variations in IQ, can nevertheless be cultivated to some extent, especially with focused training, but it would be helpful to determine the relationship between SI, maturity, and general education.

A growing body of research has focused on the construct of Spiritual Intelligence with several validated measures developed by different authors using different methods to produce a model and then validate it. The felt need for such a construct is evidenced by the emergence of more models specific to different cultures and populations, in addition to the first ones developed and initially used in cross-cultural research. Some of the SI measures have demonstrated incremental predictive validity even when controlling for other established constructs, such as personality, emotional intelligence, and social intelligence. SI has been shown to predict functioning and wellbeing in a variety of domains, producing valuable outcomes solving problems and creating products that are valuable within different cultural setting, thus meeting essential defining characteristic for intelligence. It is possible, especially now that so many related SI models exist—along with research on their predictive validity for positive outcomes compared to and controlling for EI factors—that refinement may lead to a universal SI model.

The problem remains of the mixed results within and across SI measures, raising the question whether all the measures of SI are assessing a single or slightly different constructs—a situation similar to that of EI, which is represented by several validated measures each with their own predictive validity. To further understanding of Spiritual Intelligence, it would be interesting to find the correlation between these different, independently validated SI measures to help discern to what extent they are measuring the same thing.

Clearly some of the puzzling and contradictory results found in some studies need more exploration, particularly with reference to differentiating EI and SI. It is impossible to know, at this point, which results may be a factor of cultural differences, a fault of the way some models or assessments were developed and/or translated or overlap with EI models. Behavior- or performance-based assessments of SI are conceptually challenging; however, such subjective internal phenomena as motives and traits have been successfully assessed through behavioral indicators at the level of predictive statistical validity. Some SI instruments, such as ISIS, are offered in 360-degree versions for others to rate an individual's SI based on behavioral indicators, though the extent to which self-assessment and the ratings of others are valid is not yet well established.

Supporting SI as a separate intelligence is research using different SI models that show its differential distribution across individuals, its development through training and its prediction of positive outcomes, when controlling for social intelligence, emotional intelligence, and personality, among other factors. It also seems to represent an adaptive advantage, which is supported by research in the emerging field of spiritual biology demonstrating a link between spiritual practice and the development of neuronal pathways, likely to leading to maturation and growth of the practitioner's spiritual intelligence capacities. This is matched by some evidence of a genetic component to SI. These hereditary and physiological structures, in addition to the adaptive advantage of SI for performance across a variety of domains, strongly suggest that SI is part of the human legacy evolutionarily. Indeed, this need not be too surprising, since spirituality and religion and the cultivation of their values as virtues have been a major and universal human preoccupation across all cultures and geographies, helping humans make sense of their life and their place in the cosmos, as well connect to the sacred and transcendent dimensions of their existence. In summary, in reviewing the literature on the construct of spiritual intelligence, although many questions worthy of further research remain, the growing body of research suggests that SI is deserving of being considered an intelligence.

19.2   Benefits of developing Spiritual Intelligence

As you incorporate the fact that everything and everyone is connected; that you have the innate ability to access higher sources of wisdom, you will start to connect to your intuition. Following are the benefits of spiritual intelligence, in brief:

- Being better able to manage and regulate your emotions;
- Better communication;
- Better decision-making skills;
- Improved relationships;
- Higher levels of happiness;
- More confidence;
- Better job satisfaction;
- Better leadership skills;

- Better overall perspective;
- A greater sense of inner peace, enhanced empathy and compassion, improved mental and emotional well-being, and a deeper connection to oneself and others;
- More likely to make better choices.

19.3 More Benefits of developing Spiritual Intelligence

(i) You'll be more present in the now

You will be able to establish deeper and more meaningful relationships with others, and access guidance from a higher plane every time that you need it.

(ii) You will be more resilient

You will understand that life is a benevolent path that guides you to live the higher expression of your soul. And that every challenge and setback is just a way to get you on the right track of what you were meant to live.

(iii) You will feel happier, more at peace, and confident that everyone conspires to help you win.

Knowing that love conquers all tribulations, that you are never alone, and that you came into this planet for a very important reason.

(iv) You will know exactly how to help others thrive

You will be able to connect directly to them and have a clearer sense of what they envision for themselves, and what you can do to help them live their purpose.

When you develop Spiritual Intelligence, you will be holistically attuned to the world, and you will no longer be constricted by man-made limitations.

SECTION-V

Development Techniques for Spiritual Intelligence

CHAPTER 20

DEVELOPMENT TECHNIQUES FOR SPIRITUAL INTELLIGENCE

20.1 How to develop Spiritual Intelligence?

Spiritual Intelligence is our ability to be conscious of ourselves, our environment, and our ability to tap into a higher source of wisdom. Spiritual Intelligence allows us to tap into a greater source of infinite wisdom and unconditional love. This is related to knowing that we are all connected. No matter who we are, where we are from, what our beliefs are, or how we perceive ourselves. Once we understand that everything is connected, that we are particles of a greater source of infinite love and wisdom, we can reclaim our personal responsibility for what we want to create. We shall no longer let fear stop us from taking action, as challenges will no longer seem daunting. You will understand that the Universe is abundant and unlimited. We shall know that everything that we wish is possible for us, and help is always available if we ask our higher source.

20.2 Ways to develop Spiritual Intelligence

Spirituality talks about the fact that there is another aspect to an existence apart from the physical/material realm, and you need to develop certain capacities in order to access it.

We can develop our spiritual intelligence through study and practice of skills like:

(i)   Quieting the mind,

(ii)   Tapping into our inner wisdom and intuition,

(iii)   Cultivating gratitude,

(iv)   Practicing humility and empathy,

(v)   Connecting intimately with God and others.

Here are some ways you can develop your Spiritual Intelligence:

(i) Practice Stillness

Your Spirit exists in a subtle realm of reality. It's important that you give yourself space to quiet your mind, away from all the noise and movement of the outer world, so you can connect to your Spirit.

Your soul "whispers" to you. In order to hear the messages coming from your higher self, it's important that you make room to hear.

You can practice stillness through different practices, from meditation to taking a quiet walk out in nature, to sitting below a tree. Anything that allows your mind to focus inward, instead of figuring out how to deal with the ever-changing elements of the outside world.

Stillness is a place to just be. No judgment, no rush, no guilt. Just a moment in time for you to be aware of who you are.

(ii) Cultivate Mindfulness

Mindfulness is one single faculty you can develop to improve your SQ. It can also improve your emotional intelligence and help you live your life authentically.

Start by simply noticing your thoughts and feelings without judgment throughout the day. You can set an alarm as a reminder.

(iii) Meditate

Meditation is an all-in-one tool for improving your spirituality and developing spiritual intelligence. Science has proven what Tibetan monks have known and practiced for centuries: it can help you develop Buddha's brain, an elevated state of consciousness also known as the flow state. Not only will it help you become a happier human being, but you will also become more spiritually evolved.

(iv) Develop Your Self-Awareness

This is about connecting to your true self, discovering your inner world, and living your life aligned to your soul's purpose.

We are all beautifully diverse, this is what makes our life experience so rich! The best way for you to discover your own path is to explore yourself. What are your values? What are your beliefs? What do you want to learn? What patterns do you want to let go of? What is your greater vision? What is standing in your way?

The moments of stillness will allow you to know yourself at deeper levels, but being self-aware is a practice that you should really focus on. Be aware of the sensations in your body, of the smells around you, of the task you are doing, of the thoughts that roam your mind.

Be aware of what people around you say to you (not about you, but about themselves), and most importantly, what they are not putting into words. Watch their body language, listen for what their inner world looks like, and put yourself in their shoes. By doing so you can really understand where they are at.

(v) Live Your Purpose With Intention

Intention is the why behind everything you do. It's your internal GPS system.

Knowing what you want to create and why it's important for you, will fuel your intrinsic motivation and guide your thoughts.

Live your purpose! We come to this planet to live our purpose. Our soul already knows what this purpose is, and all we have to do is remember.

Ask yourself:

- If I had all the resources that I need, what would I love to create in the world?
- What do I want to experience?
- Why is this important for me?
- What is the legacy I want to leave in the world?
- Why does this matter?

Purpose is the long-term mission of your soul.

Your intention helps you align to this purpose on a daily basis. Helping others find their purpose and consciously attune with it is key for them to be self-accountable.

(vi) Understand That Everything is Connected

Every element of creation is intertwined at a cellular level, and everything that we think, do, and say has an effect on our surroundings.

Everything is made up of energy. ***We are 4% physical and 96% energy beings.***

The tree outside your window, the cat walking down the street, the clouds, everything has energy, and we are always exchanging this energy between one another.

High-vibrational energies, such as love, empathy, compassion, and joy, help raise the energy vibration of everything around you. While low-vibrational energies, like fear, guilt, anger, or shame, do the opposite.

The outside world will always be a reflection of what is going on inside yourself. So in order to change the world, you have to change yourself.

Understanding that everything is connected also increases our personal responsibility and accountability.

If someone is triggering something in you, know that it's because there is something that you need to work on within yourself. Go deeper into that instead of rejecting the lesson.

(vii) Be Open

Open your mind to receive new information. Open your heart to give and receive unconditional love to others. Open your senses to learn about the world in different ways.

Being open means your willingness to perceive everything as it comes, not the way you think it should, but the way it is, without judgment. It's the concept of allowing your intuition to guide the way, without our intellect trying to label what comes.

When you are open, you will also learn to see beyond the surface of every situation. You will learn to view life from different perspectives. You will learn to reframe every challenge from another perspective and access different solutions.

Be open to receiving guidance and intentionally ask for it. Be open to let go of control and allow your higher self to guide the way when you need.

And be open to fully accepting where your friends are at, what their beliefs are, and the speed at which they choose to move forward. Trust that love and compassion can make a huge impact!

(viii) Read Spiritual Texts

by reading spiritual texts, you can download the universal spiritual wisdom of Buddhism, Judaism, Christianity, Hinduism, Islam, and beyond within yourself.

(ix) Cultivate Compassion

Regardless of our differences, we all strive for the same thing—to be happy, because the purpose of life is happiness, according to the Dalai Lama. And believe it or not, compassion is one of the sources of true happiness.

When you have compassion, your mind is calm because it's free from judgment and resentment. What's more, studies have shown that when you meditate on compassion, it activates your motor cortex, the area of the brain connected to the intention to act.

BIBLIOGRAPHY

Amram, Yosi. "The Seven Dimensions of Spiritual Intelligence: An Ecumenical, Grounded Theory", 2007.

Amram, Yosi. "The Intelligence of Spiritual Intelligence: Making the Case", Religions13: 1140, 2022. https:// doi.org/10.3390/ rel13121140

Amram, Yosi; and D. Christopher Dryer. "The Integrated Spiritual Intelligence Scale (ISIS): Development and Preliminary Validation", Paper Presented at the 116th Annual Conference of the American Psychological Association Boston, MA August 14-17, 2008.

Abdollahzadeh, Hasan. "The 29-item Spiritual Intelligence Questionnaire", https://www.researchgate.net/ publication/329375362, December 2009.

Bharti, Parvi. "Spiritual Quotient (SQ): Going Beyond IQ and EQ", ITM n-Ach, Vol. 7 No. 1 April 2013, pp 71-80.

Bowell, R. "The Seven Steps of Spiritual Intelligence", Nicholas Brealey Publishing, 2005

Canales, Bladimir Becerra, and Domizbeth Becerra Huaman. "Design and validation of the spiritual intelligence scale in health practice, Ica-Peru", October 2020. https://doi. org/10.6018/eglobal.417371.

Chan, A. W., & Siu, A. F. Application of the Spiritual Intelligence Self-Report Inventory (SISRI-2) Among Hong Kong University Students. International Journal of Transpersonal

Studies, 35(1), 1-12. 2016. International Journal of Transpersonal Studies, 35 (1). http://dx.doi.org/10.24972/ijts.2016.35.1.1

Daaleman, T. P. & Frey, B. B. "The Spirituality Index of Well-Being: A new instrument for health-related quality of life research", Annals of Family Medicine, 2, 499-503, 2004.

Dhar, Santosh and Upinder Dhar. "Spiritual Intelligence Scale" Agra: National Psychological Corporation, Agra. 2010.

Emmons, Robert A. and Michael E. McCullough (Eds.). "The Psychology of Gratitude", Oxford University Press, 2004.

Esmaili, Mahdi; Hamid Zareh; and Mahdi Golverdi. "Spiritual Intelligence: Aspects, Components and Guidelines to Promote it", International Journal of Management, Accounting and Economics Vol. 1, No. 2, pp. 163-174. September, 2014.

Facio, Francesca. "How To Develop Spiritual Intelligence", Evercoach Mindvalley, May 27, 2021.

Gabriel, Roger. "5 Ways To Surrender Spiritually Through Meditation", Chopra.com January 4, 2019.

Gardner, H. "Multiple intelligences: The theory in practice", Basic Books, 1993.

Goddard, Neville. "The Power of Awareness", Global Grey, 2018.

Grasmane, Ina, Vita̅lijs Rašc̆evskis, and Anita Pipere. "Primary validation of Children Spiritual Intelligence Scale in a sample of Latvian elementary school pupils" International Journal of Childrens Spirituality 27: 97–112, 2022.

Hassan, Amany Abdultawab Saleh. "The Components of the Spiritual Intelligence, Predicting the Mental Toughness and Emotional Creativity for the University Students", Hindawi

Education Research International Volume 2023, Article ID 1631978, pp. 1-13. https://doi.org/10.1155/2023/1631978.

Hildebrant, Linda S. "Spiritual Intelligence: Is it related to a leader's level of ethical development?", Ph.D. Thesis, Capella University February 2011.

Jones, G; S. Hanton, and D. Connaughton, "What is this thing called mental toughness? An investigation of elite sport performers," Journal of Applied Sport Psychology, vol. 14, no. 3, pp. 205–218, 2002.

Joshi, Amita. "Study of Spiritual Intelligence and Emotional Intelligence Related abilities of Teacher Trainees in relation to their Gender and some socio- educational factors", Thesis for Doctor of Education, Kumaun University, 2008.

Kerstetter, Bill. "Role of Spiritual Intelligence in Leader Influence on Organizational Trust", Ph.D. Thesis, Walden University, 2018.

King, D. B., and Decicco, T. L. "A viable model and self- report measure of spiritual intelligence", International Journal of Transpersonal Studies, 28, 68–85, 2009.

Kumar, V. Vineeth, and Vineeth Mehta. "Gaining adaptive orientation through spiritual and emotional intelligence". New Facets of Positivism pp. 281–301, 2011.

Mavalwalla, Jamshed M (Compiler). "Surrender In The Integral Yoga" Second Edition, Auro Publications, Sri Aurobindo Society, Puducherry, 2021.

O'Donnell, Ken. "Endo quality..." Casa da Qualidade, Brazil, 1997.

Srivastava, Prem Shankar. "Spiritual intelligence: An overview", International Journal of Multidisciplinary Research and

Development, Volume 3; Issue 3; March 2016, Page No. 224-227

Wigglesworth, C. "SQ21: The twenty-one skills of spiritual intelligence", New York, NY: SelectBooks, Inc. 2012.

Zohar, Danah. "Rewiring the Corporate Brain: Using the New Science to Rethink how we Structure and Lead", Berret Koehler, 1997.

Zohar, Danah, and Ian Marshall. "SQ-Spiritual Intelligence, the Ultimate Intelligence", Bloomsbury, London, 2000.

INDEX